CONFINED SPACE RESCUE TECHNICIAN MANUAL

—REVISED SECOND EDITION—

This manual has been approved by
the California State Fire Marshal and the
California State Board of Fire Services.

by CMC Rescue, Inc.
Goleta, California, USA

AF856336

All rights reserved. No part of this book may be reproduced, stored in a retrieval system or transmitted in any form or by any means, electronic, mechanical, photocopying, recording or otherwise without the prior written permission of the publisher.

WAIVER OF LIABILITY

The Confined Space Rescue Technician Manual is intended as an aid to training, not a substitute for it. Because we have no control over the level of training you will receive or the amount of experience you gain, neither the authors nor CMC Rescue, Inc., a California corporation, its officers, directors or shareholders are responsible for the reader's use of this book or any information contained herein.

CMC Rescue, Inc.

6740 Cortona Drive, Goleta CA 93117
Telephone: 805-562-9120 / 800-235-5741
www.cmcrescue.com

ISBN 978-0-9845875-1-3

©2017 CMC Rescue, Inc.

Revised Second Edition, October 2017

Printed in the U.S.A.

TABLE OF CONTENTS

Introduction

The fundamental goal in rescue is to remove the subject from his predicament as quickly as possible without causing further injury and with maximum safety to rescue personnel. The purpose of the *CMC Confined Space Rescue Technician Manual* is to enhance the ability of emergency responders to successfully perform a rescue from a confined space. While the material is directed toward emergency responders, it is valuable information for anyone who may be involved in a confined space emergency.

This manual is intended to support hands-on training from qualified rescue instructors who are experienced in teaching and performing confined space operations. While this manual will help responders to learn and refresh their technical knowledge, common sense and the law require regular practice and continuing education in order to maintain a responsible level of proficiency.

While we have done our best to cover the subject thoroughly, this manual cannot possibly include everything on the subject. Furthermore, the field of confined space rescue is still evolving. Responders need to stay on top of changes to regulations and standards, as well as new equipment and techniques. This manual reflects the experience we have gained teaching hundreds of rope and confined space rescue courses for more than 30 years. It has benefited from the contributions of our instructors, professional emergency responders and from thousands of students.

Confined space rescue, like other types of rescue, involves an element of risk. In this manual and in our classes, we stress the importance of safety. Safety should be an integral part of planning and training, not something practiced just during rescues. The first step is to recognize that you are responsible for your own safety and the safety of your team members. Safety is a consideration in your choice of techniques, equipment, personnel and standards. Everyone involved must be safety conscious, and it must be their first priority. All of us involved in a rescue, or even a training scenario, want to get it completed as rapidly as possible—but we must do it safely. Keep in mind that in rescue, we are not in the business of exchanging lives.

The use of the male gender throughout this text is for literary convenience and applies to rescuers of both genders. We use the term "subject" instead of "victim" as a matter of preference. When discussing medical care, we use the term "patient."

About the Revised Second Edition

After the first edition of this manual was written, the NFPA 1006 and 1670 standards were published establishing criteria for organizations and individuals responding to confined space rescue incidents. ANSI Z117.1 "Safety Requirements for Confined Spaces" was published to further influence safe entry into confined spaces. The Federal OSHA requirements were modified, which in turn caused states that developed their own requirements to change their regulations. There have also been numerous changes to equipment standards as new equipment has been developed for rope and confined space rescue. All of this needed to be reflected in this text.

Based on research conducted in CMC Rescue School classes and by others, we have modified and expanded the knots and anchors portions of Chapter 9. We have added new chapters on high point anchor systems, communications and permitting confined spaces. The glossary has been expanded to include terms used in all types of rescue and not just those that appear in the text. We have included more documents in the body of the text and reduced the size of the appendix. We continue to emphasize performance-based skills.

Where to stop has been a constant topic of discussion among our instructor staff and associates. Confined space rescue involves several other disciplines—rope, hazardous materials, medical, incident command, etc. We tried to limit the coverage of those areas to what was essential to performing a confined space rescue safely and successfully. Therefore, the rope portions are sufficient to train users to access the confined space and to extricate the subject and rescuers. But the rigging team may need additional training to perform a task, such as lowering a patient from a high-process vessel in an oil refinery. Similarly, we cover common atmospheric hazards and monitoring for confined spaces, but it would take additional hazardous materials training to detect or mitigate more exotic hazards. The readers of this book will have a wide range of medical training and experience. We have limited our coverage to packaging the patient for removal from the space and have not covered any medical treatment.

The first edition of this book was adopted by California State Fire Training as the text to accompany their 40-hour confined space rescue operations course. This edition was written concurrently with a complete revision of the instructor guide for that course, which is now called confined space rescue technician. It will continue to be the manual used for that training class throughout the state, and the inclusion of the California regulations was for that reason. The addition of California law also serves as an example of state plan regulations, but it would have been impossible to include the regulations from all of the state plan states. In the first edition we included the exact verbiage of both the California and federal regulations, but today the Internet makes that unnecessary. With a few keystrokes you can have up-to-date copies of federal, state and local regulations. We encourage you to check online for the most recent versions as these regulations will most certainly change.

CMC Editorial Staff

Rich Phillips, President and CEO, CMC Rescue, Inc.
Jim Frank, Founder and Chairman, CMC Rescue, Inc.
John McKently, Director, CMC Rescue School
Joe Flachman, CMC Rescue, Inc.
Bruce Parker, Retired, CMC Rescue School

Acknowledgements

For their valuable assistance in the writing of this book, we thank the California State Fire Training Confined Space Rescue Technician Working Group:

Kent Freeman – Captain, Roseville Fire Department, Retired
Mike Bilheimer – Captain, San Bernardino Fire Department
Wayne Chapman – Captain, Orange County Fire Authority
Stan Klopfenstein – Division Chief, Santa Fe Springs Fire Department, Retired
John McKently – Director, CMC Rescue School
Don Shellhammer – Battalion Chief, Vista Fire Department
Lou Steslicki – Division Chief, Upland Fire Department, Retired
Rodney Slaughter – Deputy State Fire Marshal, Retired

Cover and book design by Silverander Communications
Illustrations by Wes Bredall

CHAPTER ONE

CONFINED SPACE ENTRY AND RESCUE OPERATIONS

SCOPE:

This chapter serves as an introduction to confined space entry and rescue operations.

TERMINAL LEARNING OBJECTIVE:

At the conclusion of this chapter individuals will be able to list the regulations affecting confined space rescue operations, determine what a confined space and permit-required space is and discuss the reason for changes in the regulations.

ENABLING OBJECTIVES:

- List the industries exempted from permit-required confined space entry procedures
- Define a confined space
- Define a permit-required confined space
- Describe examples of permit-required confined spaces
- Describe other regulations and standards affecting confined space rescue operations
- List the causes of fatalities in confined space operations

Confined space rescue represents one of the most challenging and dangerous rescue operations undertaken by fire departments and industry today. Nearly 60% of all confined space deaths are would-be rescuers associated with secondary entries.

This manual will address the regulations governing confined space entry and rescue, rescuer safety, rescue team evaluation, preplanning, assessment and pre-entry operations, atmospheric monitoring and monitoring devices, self-contained breathing apparatus, supplied air breathing apparatus, flash protection and personal protective equipment, patient recovery systems and removal devices, lockout/tagout/blankout procedures and confined space communications equipment.

Statutes, Regulations and Standards

Entry into confined spaces has been recognized as dangerous enough that there are various federal and state laws concerning it. Federal regulations were enacted on January 14, 1993, and are contained in Vol. 58, No. 9, of the Federal Register. Chapter 29, Part 1910, Section 146 of the Code of Federal Regulations (CFR), *Permit-Required Confined Spaces*, became effective April 15, 1993. There were modifications to the regulations on June 29, 1993 (58 FR 34845), May 19, 1994 (59 FR 26115) and December 1, 1998 (63 FR 66038.) 29 CFR Part 1910 covers general industry only. Groups covered by other regulations, such as shipyard employment (Part 1915), construction (Part 1926) and agriculture (Part 1928), are not covered by this regulation. Part 1910.126 on construction was enhanced and largely follows the confined space regulations for general industry.

The federal standard is performance oriented as opposed to specification oriented. OSHA does not intend to specify exact methods of compliance. The employer's program must result in employee performance that meets the requirements of the standard and prevents unsafe acts. Just providing the training is not enough. The employer must assure that the employee can function according to the procedures the employer has developed. This has caused some concern with industry because it leaves final decisions on compliance up to the individual inspectors. Because of this, it is important to consult with your local regulatory agency on compliance.

Some states simply adopt the federal regulations, while others draft their own, which must meet or exceed the federal regulations. The California state regulations are in the *General Industry Safety Orders, Title 8 of the California Code of Regulations, Article 108, Sections 5156, 5157* and *5158*. These regulations have been amended to more closely conform with the federal regulations and became effective on December 24, 1993, with minor revisions on February 9, 1994, January 30, 1995, and July 13,1999. The California Regulations are covered in greater depth in Section 2 of this manual.

States That Have Adopted Their Own Confined Space Regulations

Alaska	Kentucky	New York*	Vermont
Arizona	Maryland	North Carolina	Virgin Islands*
California	Michigan	Oregon	Virginia
Connecticut*	Minnesota	Puerto Rico	Washington
Hawaii	Nevada	South Carolina	Wyoming
Indiana	New Jersey*	Tennessee	
Iowa	New Mexico	Utah	

*The Connecticut, New Jersey, New York and Virgin Islands plans cover public sector (state and local government) only.

The National Institute for Occupational Safety and Health (NIOSH) Publication No. 80-106, *Working in Confined Spaces*, and the American National Standards Institute (ANSI) Z117.1—*Safety Requirements for Confined Spaces*, NFPA International 1006 *Standard for Rescue Technician Professional Qualifications*, NFPA 1670 *Standard on Operations and Training for Technical Search and Rescue Incidents* have also set standards for confined space entry. Compliance with these other standards may not be required by law but is strongly recommended.

STANDARDS FOR CONFINED SPACE ENTRY AND RESCUE ADDRESS THE FOLLOWING TOPICS:

- Definitions of confined spaces and other associated terms
- Scope and application of the standard at local and national levels
- Preparation for confined space entry
- Atmospheric testing
- Attendants and rescue teams
- Permit systems
- Training
- Special equipment and tools
- Retrieval systems and respiratory protection

Other Regulations

Some other regulations that may apply while performing confined space rescues are specific to:

- Respiratory protection
- Lockout/tagout/blockout
- Fall protection
- Trench excavation

Industry Specific Regulations

Certain industries have been exempted from the general industry confined space regulation. The Federal regulation exempts three industries and in California there are eight industries listed that must abide by industry specific regulations. Section 5158 has minimum regulations that also apply, but it is important to check the industry-specific regulations because some of them are more restrictive.

When performing rescues in industries having specific regulations that only apply to them, off-site rescue teams must comply with the more stringent permit-required regulations written for the purpose of the rescue. Following the stricter regulations may appear to slow the rescue operation but in reality will assure that appropriate margins of safety are provided for the rescuers. Preplanning and practice will streamline the complex process, providing the best chance for successful rescues.

FEDERAL EXCEPTIONS TO THE STANDARD AND THE LOCATION WHERE THAT WORK IS COVERED:

- Agriculture 29 CFR 1928
- Construction 29 CFR 1926 subpart AA
- Shipyard employment 29 CFR 1915

CALIFORNIA EXCEPTIONS TO THE STANDARD AND THE LOCATION WHERE THAT WORK IS COVERED:

- CCR Title 8, Article 108, Section 1950-1952, Confined Spaces in Construction
- CCR Title 8, Article 108, Section 3437, Agricultural operations
- CCR Title 8, Article 108, Section 3460, Marine terminal operations
- CCR Title 8, Article 108, Section 8616, Telecommunications
- CCR Title 8, Article 108, Section 5178, Grain-handling facilities
- Title 49 CFR parts 191,192, 193, Natural gas utility operations within distribution and transmission vaults
- CCR Title 8, Article 108, Section 2700, Electric utility operations within underground vaults
- CCR Title 8, Article 108, Section 8355, Shipyard operations

Need for Regulation

Recent data from the Bureau of Labor Statistics (BLS) Census of Fatal Occupational Injuries (CFOI)* indicate that over the five-year period of 1997-2002, 458 fatal occupational injuries occurred in confined spaces, averaging 92 fatalities per year. Included in this study were fatalities from physical, as well as atmospheric hazards.

Of the 458 deaths during the study period, 130 workers died from being caught in, or crushed by, collapsing materials. This was the leading cause of death to workers during this study period. Second place belonged to atmospheric hazards, killing 101 workers during the study period.

It is interesting to note that this study was started in 1997, three years after the Federal Occupational Safety & Health Administration (OSHA) promulgated 1910.146. The statistics shown above support the conclusion that workers still do not fully understand the hazards involved in entering permit-required confined spaces.

*Source: *Occupational Safety and Health Magazine*. Samuel Meyer

Injury and Illness Prevention Program

By complying with the current confined space regulations, employers will be one step closer to complying with another California state regulation, CCR Title 8 § 3203, Injury and Illness Prevention Program (IIPP). This regulation was promulgated to make all aspects of the workplace safer for employees.

THE INJURY AND ILLNESS PREVENTION PROGRAM INCLUDES THE FOLLOWING POINTS:

- Safety responsibility
- Compliance/recognition
- Employer-employee communication
- Workplace inspections/evaluations
- Correction of hazards
- Injury/illness investigation
- Training
- Recordkeeping

IIPP violations are at the top of the list of citations issued by Cal/OSHA. When Cal/OSHA issues a citation they must list the regulations the employer has violated. Many unsafe acts or omissions do not fall directly under a regulation. For this reason we see many citations listing a safety responsibility violation and listing the regulation violated as the IIPP.

Some other states have similar programs.

Selected Incidents of Occupational Fatalities in Confined Spaces

Type of Space	Type of Hazard	Worker Deaths	Rescuer Deaths	Total Deaths	Comments
Sewage Digester	Oxygen Deficiency	1	1	2	—
Sewer Line Construction	Toxic Atmosphere Physical Hazard	1	1	2	38 Others Injured
Fracturing Tank	Oxygen Deficiency	0	2	2	2 Rescuers Drowned
Toluene Storage Tank	Toxic Atmosphere, Explosion, Limited Entry and Exit	1	1	2	15 Others Injured
Waste Water Tank	Toxic Atmosphere, Physical Hazard	1	1	2	Rescuer Died Two Weeks Later
"Spent" Acid Tank	Toxic Atmosphere	0	1	1	Rescuer was Father of Worker
Underground Waterline Valve Area	Toxic Atmosphere	1	0	1	Worker Died of Acute Liver Failure
Sewage Pumping Station	Physical Hazard, Toxic Atmosphere	1	3	4	2 Died of Drowning, 2 Died of Asphyxiation

The chart above shows some examples of the types of incidents investigated by OSHA during the 1980s that were the impetus for the confined space regulations to be written. Unfortunately there hasn't been a complete study of confined space rescues or fatalities done since then. As you will see from the short reports below, confined space emergencies continue to occur with disturbing regularity.

In a 2005 accident in Delaware, two workers were asphyxiated in a nitrogen-filled reactor vessel. Apparently they discovered a roll of duct tape was left in the space where they were working. They attempted to remove it by using a wire hook but that was unsuccessful. While we will never know for sure, one of the workers either entered the space in an attempt to grab the tape and get out quickly or simply slipped and fell into the space where the oxygen-depleted environment overcame him. The second worker saw his friend lying on a tray approximately five feet down inside the reactor, quickly inserted a ladder through the opening and climbed inside and also succumbed. Due to the nitrogen purge the oxygen inside of the space was less than 1%. Humans become impaired almost immediately when the oxygen level drops below about 19%. The workers were concerned that the time it would take for a properly equipped crew to enter the permit-required confined space would delay the completion of the job.

In a similar case two workers were found dead after apparently suffocating in a large stainless-steel tank in a natural foods manufacturing plant in Northern California. The tank was empty when rescue crews arrived but had been filled with argon as part of the cleaning process to expunge oxygen. Firefighters used fans to pump air into the tank before they could rescue the two men. No one knows why the men entered the tank.

Grain bins seem to be the source of many confined space fatalities. The following example from Florida resulted in $161,000 in penalties. Three workers were emptying a 42-foot-high corn silo. Two remained outside the silo while the third entered through a hatch and descended a ladder. His job was to dislodge corn that was caked inside and to keep the corn flowing to the processing area, which required that he leave the ladder. About 30 minutes later the worker, who wore no protective equipment, reported that he was sinking. The unloading machinery was stopped but by then he was already buried chest deep. His co-workers attempted to pull him out but were unsuccessful. He was finally rescued four hours later but died of asphyxiation due to the pressure of the corn on his chest. In a newspaper story about a similar incident in Minnesota an official states that on average between one and two deaths occur annually from suffocation in grain bins. In 1996 there were a record seven deaths in that state.

While any death in a confined space is tragic, the ones we really don't like to hear about are when our fellow rescuers are killed. In South Dakota there was a call to rescue a worker who had climbed into a 4,000-gallon molasses vat to clean it and had passed out. The request went out for assistance and a 17-year veteran of the local fire department, who was also an employee at the same feed mill, was one of the first rescuers at the scene. He climbed into the vat but was also overcome by toxic fumes, later determined to be hydrogen sulfide. Rescuers were unable to revive the two men and they were pronounced dead. The feed mill had been cleaning the vat the same way for the past 40 years and it wasn't clear what went wrong this time.

In Virginia methane gas emanating from a dairy farm manure pit killed five people. A farmer shimmied through a four-foot opening into the pit, which is similar to an underground tank. He climbed down a ladder and passed out. His farmhand, thinking he suffered a heart attack, followed him down a few minutes later and also passed out. Another worker notified the farmer's wife and asked her to call for help. Instead she went into the pit to help as did her two daughters. It is unsure if they suffocated from the fumes or drowned in the 18 inches of liquefied cow manure that remained in the bottom of the pit.

The point of citing these examples is not to scare you or to dwell on someone else's misfortune but to emphasize that confined spaces can be dangerous unforgiving places if you don't follow the rules. If you respond to a confined space emergency, take the time to go through all of the steps necessary to insure your safety and that of the rest of the rescue team.

Ramifications of Regulation

WHAT IS THE EFFECT OF THE OSHA STANDARD ON THE FIRE SERVICE AND ON INDUSTRY?

These standards will require industrial organizations that perform entry into confined spaces to have on location or have access to a rescue team for confined space rescue any time they enter a permit-required confined space for work. In California, Cal/OSHA has gone further than federal OSHA. In comments related to the then-proposed standards, they stated that "just calling 911 is not an adequate level of rescue and emergency protection." Outside rescue and emergency services can only be a backup to immediately available, on-site rescue services. At least one rescue trained person must be available, on-site, during all permit space entries.

THIS WILL AFFECT THE FIRE SERVICE IN SEVERAL WAYS:

- Industry will be asking for or identifying confined space rescue teams from fire departments.
- Fire departments that offer confined space rescue as a part of their technical rescue program are required to follow the standard as closely as possible. This includes having a written confined space entry and rescue program that has specialized rescue training and at least yearly refresher training.
- Revisions to the Federal Regulations require that the employer *evaluate* and *verify* that the fire department "can effectively respond in a timely manner" and that they are "equipped, trained and capable of functioning appropriately to perform permit space rescues."
- Failure to comply with the standard could result in citations.

These are the least of the problems. Confined space rescues attempted by ill-prepared and ill-equipped rescue teams not only could, but have resulted in the deaths of rescue personnel.

"An unplanned rescue will probably be your last."

Confined Spaces

Confined spaces exist in every community in a variety of forms. The regulations define a confined space as a space that:

- Is large enough and so configured that an employee can bodily enter and perform assigned work
- Has limited or restricted means for entry or exit (for example, tanks, vessels, silos, storage bins, hoppers, vaults and pits are spaces that may have limited means of entry)
- Is not designed for continuous employee occupancy

EXAMPLES OF CONFINED SPACES INCLUDE:

- Tunnels
- Sewers
- Tanks
- Tank cars and trailers
- Ship holds and accesses
- Cofferdams
- Process vessels
- Reaction vessels
- Manholes
- Industrial spaces
- Storm drains
- Ovens and furnaces

Permit-Required Confined Spaces

NIOSH studies on confined space fatalities revealed that most occurred in spaces with additional hazards and were the result of workers forgetting to perform some sort of pre-entry procedure. For that reason NIOSH recommended that spaces with certain additional hazards be subclassified, and before entry, workers be required to use a permit, or checklist, before entering the space. OSHA classifies these spaces as a **Permit-Required Confined Space**. Regulations define a **Permit-Required Confined Space** as a confined space with one or more the the following characteristics:

1. Contains, or has the potential to contain, a hazardous atmosphere
2. Contains a material that has the potential for engulfing an entrant
3. Has an internal configuration such that an entrant could be trapped or asphyxiated by inwardly converging walls or a floor that slopes downward and tapers to a smaller cross section
4. Contains any other recognized serious safety or health hazard

Other Potentially Hazardous Confined Spaces

Other spaces exist that may not meet the definition of a confined space but present many of the same hazards and may require some level of personal protection. Some examples include:

- Empty swimming pools
- Below-grade loading docks
- Parking garages
- Trenches and excavations

Confined Space Emergency

A confined space emergency is any action or event, whether inside or outside the confined space, which could endanger the persons working within the space. Emergencies include the failure of any hazard control or monitoring equipment used in the space, such as ventilators or atmospheric testers, and any unauthorized or illegal entries.

Phases of Confined Space Rescue

For our purposes, confined space rescue has been broken down into five phases.

I. Preparation

II. Assessment

III. Pre-Entry Operations

IV. Entry and Rescue Operations

V. Termination

This book will examine each of these phases and the component parts needed to effect a safe and orderly entry and rescue in a confined space.

CHAPTER TWO

COMPLIANCE WITH CALIFORNIA CODE OF REGULATIONS

General Industry Safety Orders
Title 8, Article 108
Sections 5156, 5157 & 5158

SCOPE:

This chapter serves as an introduction to the California regulations.

TERMINAL LEARNING OBJECTIVE:

At the conclusion of this chapter individuals will demonstrate a knowledge of the regulations affecting permit-required confined space entry in California.

ENABLING OBJECTIVES:

- List the regulations that pertain to confined space and permit-required confined space operations
- List the elements of the permit-required confined space entry program
- List the items necessary on an entry permit
- Describe the duties of an authorized entrant, attendant and entry supervisor
- Discuss training requirements for permit-required confined space entry personnel
- Describe the requirements for rescue and emergency services

State Regulations

To properly and safely perform confined space rescue, you must first understand the regulations. As mentioned in Chapter One the regulations may be developed and enforced in several ways. Some states leave it up to the federal government and OSHA enforces the federal regulations. Other states simply adopt and enforce the federal regulations. A third group of states, commonly referred to as "state plan states," write and enforce their own regulations, which must be equal to or more stringent than the federal regulations.

This chapter describes the California code as an example of a state that has chosen to write and enforce its own regulations. The following is a paraphrasing and explanation of the regulations. It was correct at the time this manual was published but may have changed. The most current copy of the California regulations is available at **www.dir.ca.gov**. For other state plan states check with the Department of Industrial Relations, Department of Occupational Safety and Health or similar agency.

For many years some industries chose not to comply with certain regulations. They considered OSHA fines a cost of doing business. With the increased fine structure that has been enacted, this is no longer cost efficient. In the past, fire departments have been allowed to conform to the intent of the regulation. That is no longer the case. Cal/OSHA has served notice in California that fire departments will conform to the letter of the law. Five citations were issued to a fire department that did not comply with the regulations during a confined space rescue. A number of firefighters were hospitalized as a result of this rescue.

Some other states are also requiring strict compliance. In Indiana, a fire department agreed to provide confined space rescue services for city workers without equipping and training their firefighters. IOSHA issued a 17-page listing of "serious" violations. The cover letter stated "Generally, governmental entities are not assessed monetary penalties, as IOSHA does for private sector employers. However, there is nothing in the state statutes that prohibits IOSHA from assessing penalties to public sector entities. If your agency was a private sector employer, you would have been assessed $41,600.00 in penalties for violating safe workplace standards. We do insist that all safety and health hazards be abated..."

It is important to check with your local regulatory agency to determine which regulations apply and at what level of compliance. No entry or rescue can be started until all the points in the regulation have been addressed.

THE REGULATIONS ARE DIVIDED AS FOLLOWS:

5156. Scope, Application and Definitions

a. Scope

b. Application and definitions

5157. Permit-Required Confined Spaces

a. Scope and application

b. Definitions

c. General requirements

d. Permit-required confined space program

e. Permit system

f. Entry permit

g. Training

h. Duties of authorized entrants

i. Duties of attendants

j. Duties of entry supervisors

k. Rescue and emergency services

l. Employee participation

m. Appendices A - E

5158. Other Confined Space Operations

a. Scope

b. Definitions

c. Operation procedures and employee training

d. Pre-entry

e. Confined space operations

5156. Scope, Application and Definitions

(a) Scope

This part of the standard explains that the regulations are to protect employees in general industry from the hazards of confined space entry.

(b) Application and Definitions

(1) This part states that Section 5157 shall apply for all operations and industries not covered in subsection (b)(2).

(2) This part states that Section 5158 shall apply for all operations and industries listed below (except where noted the following sections are in the California General Industry Safety Orders):

(A) **Construction operations** regulated by section 1502

(B) **Agriculture operations** (including cotton gins) defined by section 3437

(C) **Marine terminal operations** defined in section 3460

(D) **Shipyard operations** regulated by section 8355

(E) **Telecommunications manholes and unvented vaults** regulated by section 8616

(F) **Grain handling facilities** regulated by section 5178

(G) **Natural gas utility operation** within distribution and transmission facility vaults defined in Title 49 Code of Federal Regulations Parts 191, 192 and 193;

(H) **Electrical utility operations** within underground vaults. See section 2700 for a definition of vault.

5157. Permit-Required Confined Spaces

(a) Scope and Application

This part explains that these requirements for practices and procedures are to protect employees from the hazards of entry into permit-required confined spaces. They apply to employers as specified in section 5156(b)(1), which applies to all industries not specified in section 5156(b)(2).

(b) Definitions

This part gives definitions for terms used in the standard. Some of the more relevant definitions include:

Confined Space means a space that:

(1) Is large enough and so configured that an employee can bodily enter and perform assigned work

(2) Has limited or restricted means for entry or exit (for example, tanks, vessels, silos, storage bins, hoppers, vaults and pits are spaces that may have limited means of entry.)

(3) Is not designed for continuous human occupancy

Permit-Required Confined Space (permit space): a confined space that has one or more of the following characteristics:

(1) Contains or has a potential to contain a hazardous atmosphere

(2) Contains a material that has the potential for engulfing an entrant

(3) Has an internal configuration such that an entrant could be trapped or asphyxiated by inwardly converging walls or by a floor which slopes downward and tapers to a smaller cross-section

(4) Contains any other recognized serious safety or health hazard

Non-Permit-Required Confined Space: a confined space that does not contain or, with respect to atmospheric hazards, have the potential to contain, any hazards capable of causing death or serious physical harm.

Entry: the action by which a person passes through an opening into a permit-required confined space. Entry includes ensuing work activities in that space and is considered to have occurred as soon as any part of the entrant's body breaks the plane of the opening into the space.

Entry Permit: the worksheet or documentation form used to list all actual or potential hazards inherent to a particular space; the procedures necessary to protect an entrant from such hazards; and the list of personnel in required positions. (See also page 2-12).

For additional definitions, see the glossary of this manual.

(c) General Requirements

(1) Employers must conduct a survey of their business site to determine whether permit-required confined spaces exist.

- Cal/OSHA provides a decision flow chart (Appendix A of the standard) to assist you in this determination. Using a confined space evaluation form, like the one shown in the appendix, is also an effective way to complete this task.
- Fire department personnel should evaluate the possible entry sites in their response area.
- The confined spaces must be evaluated for the type of hazards present. These spaces are then classified as permit-required or non-permit-required spaces, as defined in subsection (b).

(2) Employees must be informed of the existence, location and danger of these spaces by signs or other effective means. Other effective means may include training or other employee information programs.

(3) If it is not necessary for employees to ever enter the confined spaces, effective measures must be taken to prevent employees from entering them.

- Warning signs may qualify but permanent barriers are the safest way to prevent entry. If periodic inspection from outside is necessary, locked covers with only qualified inspectors having key access will prevent unauthorized entry.

Example of a confined space warning sign

(4) If employees will enter permit spaces, a written permit-required confined space entry program must be developed and implemented. This written program must be made available for inspection by employees and their representatives.

- Cal/OSHA gives specific guidance for establishing this written program in subsection (d). Sample programs that comply are also provided in Appendix C of the standard.

(5) An employer may use the alternate procedures specified in (c)(5)(B) if the conditions specified in (c)(5)(A) are present.

- These alternate entry procedures require prior testing and documentation and are of little use to rescue teams usually responding to a rescue where the ventilation has failed.

(A) Entry into a permit space need not comply with (d) permit-required confined space program, (e) permit system, (f) entry permit, (h) duties of authorized entrants, (i) duties of attendants, (j) duties of entry supervisors and (k) rescue and emergency services provided that:

1. The employer can demonstrate that the only hazard is an actual or potential hazardous atmosphere.
2. The employer can demonstrate that continuous forced-air ventilation alone is sufficient to maintain the permit space safe for entry.
3. The employer develops monitoring and inspection data that supports (c)(5)(A)1. and 2.
4. If initial entry is necessary to obtain the data for (c)(5)(A)3., the entry is made as a permit-required confined space entry.
5. The determinations and supporting data required by (c)(5)(A)1., 2. and 3. are documented by the employer and made available to each employee who enters the confined space.
6. Entry into the permit space is performed in accordance with (c)(5)(B).

(B) The following requirements apply to entry into permit spaces that meet the conditions set forth in (c)(5)(A):

1. Any conditions that make it unsafe to remove an entrance cover shall be eliminated before removing the cover.
2. When entrance covers are removed, the opening shall be promptly guarded by a railing, temporary cover, or other temporary barrier that will prevent an accidental fall through the opening and that will protect each employee working in the space from foreign objects entering the space.
3. Before an employee enters the space, the internal atmosphere shall be tested with a calibrated direct-reading instrument for the following conditions in the order given:

   a. Oxygen content

   b. Flammable gases and vapors

   c. Potential toxic air contaminants

4. No hazardous atmosphere may be within the space whenever an employee is in the space.
5. Continuous forced air ventilation shall be used as follows:
   a. An employee may not enter the space until the forced-air ventilation has eliminated any hazardous atmosphere.
   b. The forced-air ventilation shall be directed as to ventilate the immediate areas where an employee is or will be present within the space and shall continue until all employees have left the space.
   c. The air supply for the forced-air ventilation shall be from a clean source and may not increase the hazards in the space.
6. The atmosphere within the space shall be periodically tested as necessary to ensure that the forced air ventilation prevents the accumulation of a hazardous atmosphere.
7. If a hazardous atmosphere is detected during entry:
   a. Each employee shall leave the space immediately.
   b. The space shall be evaluated to determine how the hazardous atmosphere developed.
   c. Measures shall be implemented to protect employees from the hazardous atmosphere before a subsequent entry takes place.
8. The employer shall verify through a written certification that contains the date, location of the space, and the signature of the person providing the certification, that the space is safe for entry and that the measures required by (c)(5)(B) have been taken.
   - The certification shall be made before entry and made available to each employee entering the space.

(6) When the use or configuration of a non-permit confined space changes, potentially increasing its hazards, reevaluate and, if necessary, reclassify it as a permit-required confined space.

(7) Under certain conditions, a permit-required confined space may be reclassified as a non-permit-required confined space by following these procedures:

(A) If the permit space has no actual or potential atmospheric hazards and if all hazards within the space are eliminated without entry into the space, the permit space may be reclassified as a non-permit confined space for as long as the non-atmospheric hazards remain eliminated.

(B) If it is necessary to enter the permit space to eliminate hazards, such entry shall be a permit-required confined space entry. If testing and inspection during entry demonstrate that the hazards within the permit space have been eliminated, the permit space may also be reclassified as a non-permit confined space for as long as the hazards remain eliminated.

- Control of atmospheric hazards through forced-air ventilation does not constitute elimination of hazards. This is covered in subsection (c)(5).

(C) The employer shall document the basis for determining that all hazards have been eliminated through a certification that contains:

- The date
- The location of the space
- The signature of the person making the determination
- The certification shall be made available to each employee entering the space

(D) If hazards arise within a permit space that has been reclassified to a non-permit space under (c)(7), each employee shall exit the space

- The employer must then reevaluate the space to determine whether it must be reclassified as a permit space.

(8) If an independent contractor must enter permit spaces, prior to entry, the host employer must inform the contractor of:

(A) The permit program requirements

(B) The permit space hazards

(C) The safety precautions and procedures that have been implemented by the host employer

(D) If employees of both the host employer and the contractor will be working in or near the same permit space, the employer must coordinate entry operations with the contractor as required by subsection (d)(11)

(E) Following the entry the contractor must be debriefed on the entry operation

(9) In addition to complying with permit space requirements that apply to all employers, each outside contractor who performs permit space entry operations shall:

(A) Obtain available information on hazards and entry operations from the host employer.

(B) If employees of both the host employer and the contractor will be working in or near the same permit space, the contractor must coordinate entry operations with the employer.

(C) Inform the host employer of the permit program followed and the hazards confronted or created during the entry operation.

(d) Permit-Required Confined Space Programs Require Employers to:

(1) Implement measures to prevent unauthorized entry.

(2) Identify and evaluate hazards before entry.

(3) Develop and implement procedures and practices necessary for safe entry operations that include at least the following:

(A) Specify acceptable entry conditions.

(B) Isolate the permit space.

(C) Purging, inerting, flushing or ventilating the permit space as necessary to eliminate or control atmospheric hazards.

(D) Provide pedestrian, vehicle, or other barriers as necessary to protect entrants from external hazards.

(E) Verifying that conditions are acceptable for the duration of the entry.

(4) At no cost to employees, provide, maintain and ensure proper training and use of the following equipment:

(A) Testing and monitoring equipment needed to comply with subsection (d)(5)

(B) Ventilating equipment

(C) Communication equipment necessary for compliance with subsections (h)(3) and (i)(5)

(D) Personal protective equipment as needed to protect employees

(E) Lighting equipment needed for safe work and emergency exit

(F) Barriers and shields as required by subsection (d)(3)(D)

(G) Equipment, such as ladders, needed for safe ingress and egress

(H) Rescue and emergency equipment as needed to comply with subsection (d)(9)

(I) Any other equipment needed for safe entry into and rescue from permit spaces

(5) Evaluate permit space conditions when entry operations are conducted.

(A) Before entry is authorized to begin.

- If isolation of the space is infeasible because the space is large or is part of a continuous system (such as a sewer), pre-entry testing shall be done to the extent feasible and entry conditions shall be continuously monitored in work areas.

(B) Test or monitor as necessary to determine that acceptable entry conditions are being maintained during the course of the entry.

(C) When testing for atmospheric hazards, test in the following order:

- Oxygen
- Combustible gases or vapors
- Toxic gases or vapors

(6) Provide at least one attendant outside the permit space for the duration of entry operations.

- Attendants may monitor more than one space as long as they can perform their duties effectively.
- Attendants may be stationed at any location outside the space as long as they can perform their duties effectively.

(7) If multiple spaces are to be monitored by one attendant, provide the means for the attendant to respond to an emergency at one or more of the entry operations without distraction from the attendant's responsibilities at other entry operations.

(8) Designate the persons who have active roles in entry operations, identify their duties and provide training for them as required by subsection (g).

(9) Develop and implement procedures for rescuing entrants from permit spaces, for providing necessary emergency services to rescued employees, for summoning additional rescue and emergency services and for preventing unauthorized personnel from attempting a rescue.

(10) Develop and implement a system for preparation, issuance, use and cancellation of entry permits.

(11) Develop and implement procedures to coordinate entry operations when employees of more than one employer are working simultaneously as authorized entrants.

(12) Develop and implement procedures necessary for concluding the entry after entry operations have been completed.

(13) Review and revise the permit space program as necessary to correct deficiencies whenever the employer has reason to believe the employees are not protected. This must happen before any subsequent entries.

Circumstances requiring this review are:

- Any unauthorized entry of a permit space
- Detection of a hazard not covered by the permit
- Detection of a condition not covered by the permit
- The occurrence of an injury or near-miss during an entry
- A change in the use or configuration of a permit space; and/or
- Employee complaints about the effectiveness of the program

(14) Review the permit-required confined space program, using the canceled permits, within one year after each entry and revise as necessary to ensure that employees are protected from permit space hazards.

  - Employers may perform one annual review covering all entries performed during a 12-month period.

(e) Permit System

(1) Before entry is authorized, the employer shall document completion of the measures necessary to make the permit space safe by filling out an entry permit.

(2) Before entry begins, the entry permit must be signed by the entry supervisor identified on the permit.

(3) The completed entry permit shall be posted at the entry portal or otherwise be made available for the entrants to see that pre-entry preparations have been completed.

(4) The duration of the permit must not exceed the time required to complete the task listed on the permit.

(5) The entry supervisor shall terminate the entry and cancel the permit when:

(A) The entry operations covered by the permit have been completed.

(B) A condition not allowed by the permit arises in or near the space.

(6) The employer shall retain the canceled permit for at least one year to facilitate the review of the permit space required by subsection (d)(14).

- Any problems encountered during the entry shall be noted on the permit so appropriate revisions can be made to the permit space program.

(f) Entry Permit

The entry permit documents that everything required by the permit system is completed before any permit-required confined space entry. The standard gives an example of a confined space entry permit in Appendix D, but any format is acceptable that includes the following information:

(1) The permit space to be entered

(2) The purpose of the entry

(3) Date and duration of the permit

(4) The names of the authorized entrants and a system to track them for the duration of the permit

- A reference to the roster or tracking system used is sufficient.

(5) The name of the attendant(s)

(6) The name of the entry supervisor(s), with the signature or initials of the entry supervisor who originally authorized entry

(7) List of hazards of the space

(8) List of hazard controls necessary before entry

(9) List of acceptable entry conditions

(10) Recorded test results with the time tested and the names or initials of the testers

(11) Rescue and emergency services that are provided on site

(12) Communication procedures for use by attendants and entrants

(13) All equipment needed for the entry

(14) Any other information specific to the entry

(15) Any other permits necessary for the work authorized (such as hot work permits)

(g) Training

The employer is responsible for providing the training and certification mandated in this part of the regulation.

(1) The employer must provide the training necessary for employees to safely perform their duties that deal with confined space entry.

(2) Each affected employee shall be trained:

(A) Before assigned to perform confined space entry.

(B) Before there is a change in assigned duties.

(C) Whenever the operation changes and the employee is not trained in the new procedures.

(D) Whenever the employer has reason to believe that the permit-required confined space program is not being followed or that the employee is not adequately trained in the procedures.

(3) The specific training required is not listed, but the training must establish employee proficiency in all duties and include new or revised procedures.

(4) The employer must certify that the training has been accomplished. This certification must include the employee's name, the signatures or initials of the trainers and the dates of the training.

(h) Duties of Authorized Entrants

An authorized entrant is an employee who is designated by the employer to enter a permit space. It is the employer's responsibility to ensure that all Authorized Entrants receive the appropriate training and perform their assigned duties properly as designated under the permit space program.

(1) Know the hazards that may be faced during the entry.

- Employers must ensure that authorized entrants know the hazards of the entry. This includes the mode of exposure to the hazards, including inhalation into the lungs, ingestion through the mouth, or absorption through the skin.
- Recognize the signs and symptoms of hazard exposure.
- Authorized entrants must recognize the effects of exposure, particularly the early warning signs and symptoms, so self-rescue can start before outside help is needed.
- Understand the consequences of hazard exposure.

(2) Use equipment properly.

- Authorized entrants must be provided with the necessary Personal Protective Equipment (PPE) and other required equipment, listed in (d)(4) of the standard, and trained to properly use it. It is the responsibility of the employer to assure that it is used.

(3) Communicate with the attendant.

- Authorized entrants must maintain contact with the Attendant. This improves their chances of safe exit. Two-way radios or other continuous electronic monitoring equipment, in combination with voice or alarms, is considered effective means of communication. This enables the Attendant to monitor the status of the Authorized Entrants, especially subtle changes in speech or deviation from set communication procedures. The Attendant can alert the authorized entrants to evacuate the space if needed.

(4) Alert the attendant of hazards that arise during the entry

(A) Whenever an authorized entrant recognizes any warning signs or symptoms of exposure

(B) Detects a condition prohibited by the entry permit

- The attendant can then alert the other authorized entrants. This allows for an organized evacuation.

(5) Exit the permit space quickly whenever:

(A) An evacuation order is given by the attendant or entry supervisor

(B) The authorized entrant recognizes the signs or symptoms of exposure to a dangerous situation

(C) The authorized entrant detects a prohibited condition

(D) An evacuation alarm is sounded

(i) Duties of Attendants

Attendants are stationed outside one or more permit spaces to monitor the condition and location of authorized entrants and perform the other duties listed in the employer's permit space program. At least one trained attendant is on duty at all permit entries. Their duties include:

(1) Know the hazards that may be faced during the entry.

- Employers must ensure that the attendant knows the hazards of the entry into the permit space. This includes the modes, signs and symptoms, and consequences of exposure.

(2) Know the behavioral effects of exposure.

- Knowing the behavioral effects of hazard exposure gives the attendant the judgment to order the evacuation of entrants if any of the signs or symptoms are detected.

(3) Maintain accurate entrant identification.

- Maintaining an accurate entrant count ensures that the means used to identify entrants under (f)(4) of the standard can accurately identify those in the permit space. By keeping an accurate count no one will be accidentally left in the permit space, or, in case of emergency, no useless searches are done to look for someone not in the permit space.

(4) Remain outside the permit space.

- Keeping unauthorized individuals out of the permit space, staying alert to hazards, communicating with entrants and summoning rescue services if needed are duties requiring the attendant to remain outside the permit space. Some permit space programs allow the attendant to enter the space for rescue, but only after being replaced by another trained attendant.

(5) Communicate with the entrants.

- Attendants must maintain effective communication with all authorized entrants during the entire entry. This allows the attendant to monitor entrant status and alert them of the need to evacuate if necessary.

(6) Monitor entry activities.

- The attendant monitors activity both inside the permit space and in the immediate vicinity outside the permit space to determine if it is safe for entrants to remain in the permit space. He orders evacuation whenever he:

  (A) Detects a prohibited condition

  (B) Detects behavioral effects of exposure in an entrant

  (C) Detects a situation outside the space that could endanger the entrants

  (D) Cannot effectively perform his duties

- If the attendant is monitoring more than one entry and has an emergency in one of the them that requires all his attention, he must terminate the other entries.

(7) Initiate on-site rescue procedures and, if necessary, summon rescue and emergency services.

- As soon as the attendant feels that an entrant may need help to exit a permit space, he should immediately begin any non-entry rescue procedures. Making contact with the rescuers he has summoned and briefing them on the hazards and subject information when they arrive are important duties of the attendant.

(8) Prevent unauthorized entry by:

(A) Warning others to stay away from the permit space.

(B) Advising unauthorized entrants to exit if they have entered the permit space.

(C) Informing the authorized entrants and the Entry Supervisor that unauthorized persons have entered the permit space.

(9) Perform non-entry rescues.

- The attendant must have the skills and knowledge to perform non-entry rescues as specified in the permit entry program.

(10) Perform no conflicting duties while acting as the attendant.

- Attendants may perform other duties, such as passing equipment into the space, only if it does not interfere with his primary duties as attendant. If the attendant is monitoring entry into more than one permit space and must focus on a rescue in one of them, he should order evacuation of the other space.

(j) Duties of Entry Supervisors

An entry supervisor is anyone empowered by the employer to authorize or directly supervise entry operations in a permit space.

(1) Know the hazards that may be faced during entry.

- The modes of exposure
- The signs and symptoms of exposure
- The consequences of exposure

(2) Verify that the appropriate entries have been made on the permit, that all tests specified by the permit have been completed and that all procedures and equipment required by the permit are in place before endorsing the permit and allowing entry to begin.

(3) Terminate entry as required by subsection (e)(5).

- When acceptable entry conditions are not present, the entry supervisor must terminate the entry operations and cancel the authorization to enter until acceptable conditions are present.

(4) Verify that rescue services are readily available and the means for summoning them are operable.

- Whether on-site or off-site rescue services are used, they must be available in time to effect a rescue and the means for requesting them should be tested.

(5) Remove unauthorized individuals who enter or try to enter the permit space during entry operations.

- Unauthorized individuals may endanger themselves, authorized entrants, or rescue personnel called to rescue them. The entry supervisor has the authority to take appropriate action to deny entry to these individuals.

(6) Determine that entry and work operations remain consistent with the entry permit terms and that acceptable entry conditions are maintained.

- Conditions may change during entry and work in permit spaces. The entry supervisor must determine that acceptable conditions are maintained. If the responsibility for a permit space entry is transferred to another person, the outgoing entry supervisor must determine that acceptable conditions still exist before transferring operational responsibility to the incoming entry supervisor.

(k) Rescue and Emergency Services

The employer shall ensure that at least one stand-by person at the site is trained and immediately available to perform rescue and emergency services.

(1) Employers who have employees perform rescue services must ensure that:

(A) Each member of the rescue service must be provided with, and trained to use properly, the PPE and rescue equipment necessary for making rescues from permit spaces.

(B) Each member of the rescue service shall be trained to perform the assigned rescue duties and receive the training required of authorized entrants.

(C) Each member of the rescue service shall practice making permit space rescues at least once every 12 months.

- These simulated rescue operations may involve dummies, manikins, or actual people being removed from the actual permit spaces or from representative permit spaces.

- Representative permit spaces shall simulate the opening size, configuration, and accessibility of the permit spaces from which rescue would be performed.

(D) Each member of the rescue service shall have training in basic first aid and cardiopulmonary resuscitation (CPR).

- At least one member holding current certification in first aid and CPR shall be available.

(2) When the employer arranges to have contract rescue services perform rescues on his site, the employer shall:

(A) Inform the rescue service of the hazards they may confront while performing the permit space rescue.

(B) Provide the rescue service with access to all permit spaces from which permit space rescue may be necessary.

- This allows the rescue service to develop appropriate pre-plans and practice rescue operations.

(3) To facilitate non-entry rescues, retrieval systems or other methods shall be used whenever an entry is made into a permit space, unless the retrieval system would increase the overall risk of the entry or would not contribute to the rescue.

Retrieval systems must meet the following requirements:

(A) Each entrant shall use a full-body harness, with a retrieval line attached at a suitable point so that when rescued, the entrant presents the smallest possible profile.

- For example, attach the line to the center of the entrants back or above the entrant's head.
- Wristlets may be used if the employer can demonstrate the body harness is infeasible or creates a greater hazard.

(B) The other end of the retrieval line shall be attached to a mechanical device or a fixed point outside the permit space in such a manner that rescue can begin as soon as the rescuer becomes aware that rescue is necessary.

- A mechanical device shall be available to retrieve personnel from vertical type permit spaces more than five feet deep.

(4) If an entrant is exposed to a substance for which a material safety data sheet (MSDS) is required to be kept at the worksite, that MSDS or written information shall be made available to the medical facility treating the exposed entrant.

(l) Employee Participation

(1) Employers shall consult with affected employees and their authorized representatives on the development and implementation of all aspects of the permit space program required by subsection (c).

(2) Employers shall make available to affected employees and their authorized representatives all information required to be developed by this section.

(m) Appendices

Appendices A–E serve to provide information and non-mandatory guidelines to assist employers in complying with the appropriate requirements of this section.

A. Permit-Required Confined Space Decision Flow Chart

B. Procedures for Atmospheric Testing

C. Examples of Permit-Required Confined Space Programs

D. Confined Space Entry Permit

E. Sewer System Options

5158. Other Confined Space Operations

(a) Scope

For industries and operations listed in section 5156(b)(2) this section prescribes minimum standards for preventing employee exposure to dangerous air contamination and/or oxygen deficiency in confined spaces.

- It is important to refer to the sections that regulate the specific industries because some have more restrictive requirements.

(b) Definitions

Definitions for this section are different than Section 5157.

(1) **Confined Space.** A space defined by the concurrent existence of the following conditions:

(A) Existing ventilation is insufficient to remove dangerous air contamination and/or an oxygen deficiency which may exist or develop.

(B) Ready access or egress for the removal of a suddenly disabled employee is difficult due to the location and/or size of the opening(s).

(2) **Dangerous Air Contamination.** An atmosphere presenting a threat of causing death, injury, acute illness or disablement due to the presence of flammable and/or explosive, toxic and otherwise injurious or incapacitating substances.

(A) Dangerous air contamination due to the flammability of a gas or vapor is defined as an atmosphere containing the gas or vapor at a concentration greater than 20% of its lower explosive (lower flammable) limit.

(B) Dangerous air contamination due to a combustible particulate is defined as a concentration greater than 20% of the minimum explosive concentration of the particulate.

(C) Dangerous air contamination due to the toxicity of a substance is defined as the atmospheric concentration immediately hazardous to life or health.

- This definition does not preclude the requirement to control harmful exposures, under provisions of Article 107, to toxic substances at concentrations less than those immediately hazardous to life or health.

(3) **Oxygen Deficiency.** An atmosphere containing oxygen at a concentration of less than 19.5% by volume.

(c) Operation Procedures and Employee Training

(1) **Operating Procedures**

(A) Written, understandable operating and rescue procedures shall be developed and provided to affected employees.

(B) Operating procedures shall conform to applicable requirements and provide for surveillance of the surrounding area to avoid hazards such as drifting vapors.

(2) **Employee Training:** Employees, including stand-by persons, shall be trained in the operating and rescue procedures, including instructions as to the hazards they may encounter.

(d) Pre-entry

(1) Lines which may convey flammable, injurious, or incapacitating substances into the space shall be disconnected, blinded, or blocked off by other positive means to prevent the development of dangerous air contamination and/or oxygen deficiency within the space.

- This does not require blocking of all laterals to sewers and storm drains. Where experience or knowledge of industrial use indicates materials resulting in dangerous air contamination may be dumped into an occupied sewer, all such laterals shall be blocked.

(2) The space shall be emptied, flushed or otherwise purged of flammable, injurious or incapacitating substances to the extent feasible.

(3) The air shall be tested with an appropriate device or method to determine whether dangerous air contamination and/or oxygen deficiency exists, and a written record of such testing results shall be made and kept at the work site for the duration of the work.

- Affected employees and/or their representatives shall be afforded the opportunity to review and record the test results.

(4) Where interconnected spaces are blinded off as a unit, each space shall be tested and the results recorded.

- The most hazardous condition found shall govern the procedures to be followed.

(5) If the tests show that dangerous air contamination and/or oxygen deficiency does not exist within the space, entry and work within the space may proceed subject to the following provisions:

(A) Testing shall be conducted with sufficient frequency to ensure that dangerous air contamination and/or oxygen deficiency does not occur during the performance of the operation.

(B) If the development of these hazards is imminent, the requirements prescribed by subsection (e) shall also apply.

(6) Where the existence of these atmospheric hazards is demonstrated by tests, existing ventilation shall be augmented by appropriate means.

(7) When subsequent testing shows that the additional ventilation has removed the atmospheric hazards, entry and work within the space may proceed subject to provisions in subsection (d)(5).

(8) Do not introduce any source of ignition until testing has determined that air contamination by a flammable or explosive substance does not exist.

(9) Whenever oxygen-consuming equipment, such as heaters, plumber's torches or furnaces, are to be used, measures shall be taken to ensure adequate combustion air and exhaust gas venting.

(10) To the extent feasible, provisions shall be made to permit ready entry and exit.

(11) Where it is not feasible to provide ready exit from spaces equipped with automatic fire suppression systems employing harmful design concentrations of toxic or oxygen-displacing gases or total foam flooding, such systems shall be deactivated.

- Where it is not practical or safe to deactivate these systems, the provisions of subsection (e) related to the use of respiratory protective equipment shall apply.

(e) Confined Space Operations

(1) Entry Into and Work Within Confined Spaces: The requirements of this section apply to entry into and work within a confined space whenever an atmosphere free of dangerous air contamination and/or oxygen deficiency cannot be ensured through implementation of:

- Ventilation
- Blinding, blocking and disconnecting of lines
- Emptying, flushing or purging

or whenever, due to the existence of an emergency, it is not feasible to ensure the removal of dangerous air contamination.

(A) Tanks, vessels or other confined spaces with side and top openings shall be entered from side openings when practicable.

- Side openings are those within 3½ feet of the bottom.

(B) Appropriate, approved respiratory protective equipment, in accordance with section 5144, shall be provided and worn.

(C) An approved safety belt with an attached line shall be used.

- The free end of the line shall be secured outside the entry opening.
- The line shall be at least 1/2-inch diameter and 2,000 pounds test.
- Exception: Where it can be shown that a safety belt and attached line would further endanger the life of the employee.

(D) At least one employee shall stand by on the outside of the confined space ready to give assistance in case of emergency. At least one additional employee who may have other duties shall be within sight or call of the stand-by employee.

1. The stand-by employee shall have respiratory protective equipment, including an independent source of breathing air, available for immediate use.
2. A stand-by employee (or employees) with respiratory protection may enter the confined space but only in case of emergency and only after alerting at least one additional employee outside the confined space of the emergency and his intent to enter the confined space.

(E) When entry must be made through the top opening, the following requirements also apply:

1. The safety belt shall be of a harness type that suspends the person in an upright position.

2. A hoisting device or other effective means shall be provided for lifting employees out of the space.

(F) Work involving flame, arc, spark or other source of ignition is prohibited within the confined space (or any adjacent space having common walls, floor or ceiling with the confined space) which contains, or is likely to develop, dangerous air contamination due to flammable and/or explosive substances.

(G) Whenever inert gases are used for preventing ignition of flammable gases or vapors, no ignition source shall be permitted unless the oxygen concentration is maintained at less than 20% of the concentration that will support combustion.

1. Testing will be done with sufficient frequency to ensure conformity.
2. A written record of the test results will be made and kept at the work site for the duration of the work.
3. Affected employees shall be provided with an opportunity to review and record the results.

(H) Only approved lighting and electrical equipment, in accordance with the Low-Voltage Electrical Safety Orders, shall be used in confined spaces subject to dangerous air contamination by flammable and/or explosive substances.

(I) Employees working in confined spaces which have last contained substances corrosive to the skin or substances which can be absorbed through the skin shall be provided with, and shall be required to wear, appropriate personal protective clothing or devices in accordance with Article 10.

(2) Precautions for Emergencies Involving Work in Confined Spaces.

(A) At least one person trained in first aid and cardiopulmonary resuscitation (CPR) shall be immediately available whenever use of respiratory protective equipment is required. (AHA or ARC)

(B) An effective means of communication between employees inside a confined space and a stand-by employee shall be provided and used whenever respiratory protection is used or whenever inside the confined space are out of sight of the stand-by employee.

- All affected employees shall be trained in the use of such communication equipment.
- The communication system shall be tested before each use to ensure its effective operation.

The most current copy of the California regulations is available at www.dir.ca.gov. For other state plan states check with the Department of Industrial Relations, Department of Occupational Safety and Health or similar agency.

CHAPTER THREE

COMPLIANCE WITH CODE OF FEDERAL REGULATIONS

Permit-Required Confined Spaces
29 CFR 1910.146

SCOPE:

This chapter serves as an introduction to the federal regulations.

TERMINAL LEARNING OBJECTIVE:

At the conclusion of this chapter individuals will demonstrate a knowledge of the Federal Regulations affecting confined space entry and rescue.

ENABLING OBJECTIVES:

- List the regulations that pertain to confined space and permit-required confined space operations
- List the elements of the permit-required space entry program
- List the items necessary on an entry permit
- Describe the duties of an authorized entrant, attendant and entry supervisor
- Discuss the training requirements for permit-required confined space entry personnel
- Describe the requirements for rescue and emergency services

The following regulation has been adopted by the federal government.

THE REGULATION IS DIVIDED AS FOLLOWS:

1910.146 Permit-required confined spaces

a. Scope and application
b. Definitions
c. General requirements
d. Permit-required confined space program
e. Permit system
f. Entry permit
g. Training
h. Duties of authorized entrants
i. Duties of attendants
j. Duties of entry supervisors
k. Rescue and emergency services
l. Employee participation

Appendices A–F

1910.146 Scope, Definition and General Requirements

(a) Scope

This part of the standard explains that the regulations are to protect employees in general industry from the hazards of entry into permit-required confined spaces. It also excludes specific industries that are covered by other regulations. They are:

- Agriculture (Part 1928)
- Construction (Part 1926 Sub Part AA) Now essentially follows the same regulations as for General Industry
- Shipyard employment (Part 1915)

(b) Definitions

This part gives definitions for terms used in the standard. Some of the most relevant definitions follow.

Confined space is a space that:

(1) Is large enough and so configured that an employee can bodily enter and perform assigned work; and

(2) Has limited or restricted means for entry or exit; and

(3) Is not designed for continuous human occupancy.

Permit-Required Confined Space (permit space) means a confined space that has one or more of the following characteristics:

(1) Contains or has a potential to contain a hazardous atmosphere;

(2) Contains a material that has the potential for engulfing an entrant;

(3) Has an internal configuration such that an entrant could be trapped or asphyxiated by inwardly converging walls or by a floor which slopes downward and tapers to a smaller cross section; or

(4) Contains any other recognized serious safety or health hazard.

Non-Permit-Required Confined Space means a confined space that does not contain or, with respect to atmospheric hazards, have the potential to contain any hazards capable of causing death or serious physical harm.

Entry means the action by which a person passes through an opening into a permit-required confined space. Entry includes ensuing work activities in that space and is considered to have occurred as soon as any part of the entrant's body breaks the plane of an opening into the space.

(c) General Requirements

(1) Employers must conduct a survey of their business site to determine whether permit-required confined spaces exist.

- OSHA provides a decision flow chart (Appendix A of the standard) to help you. Using a confined space evaluation form, like the one shown in the back of this manual, is also an effective way to complete this task.
- Fire department personnel must evaluate the possible entry sites in their response area.
- The confined spaces must be evaluated for the type of hazards present. These spaces are then classified as permit-required or non-permit-required spaces, as defined in subsection (b).

(2) Employees must be informed of the existence, location and danger of these spaces by signs or other effective means. Other effective means may include training or other employee information programs.

(3) If it is not necessary for employees to ever enter the confined spaces, effective measures must be taken to prevent employees from entering them.

- Warning signs may qualify but permanent barriers are the safest way to prevent entry. If periodic inspection from outside is necessary, locked covers with only qualified inspectors having key access will prevent unauthorized entry.

(4) If employees will enter permit spaces, a written permit-required confined space entry program must be developed and implemented. This written program must be made available for inspection by employees and their representatives.

- OSHA gives specific guidance for establishing this written program in subsection (d). Sample programs that comply are also provided in Appendix C of the standard.

(5) An employer may use the Alternate Procedures specified in (c)(5)(ii) if the conditions specified in (c)(5)(i) are present.

- These alternate entry procedures require prior testing and documentation and are of little use to rescue teams usually responding to a rescue where the hazard controls have failed.

(i) Entry into a permit space need not comply with (d) Permit-required confined space program, (e) Permit system, (f) Entry permit, (h) Duties of authorized entrants, (i) Duties of attendants, (j) Duties of entry supervisors, and (k) Rescue and emergency services, provided that:

(A) The employer can demonstrate that the only hazard is an actual or potential hazardous atmosphere;

(B) The employer can demonstrate that continuous forced-air ventilation alone is sufficient to maintain that permit space safe for entry;

(C) The employer develops monitoring and inspection data that supports (c)(5)(i)(A) and (B);

(D) If initial entry is necessary to obtain the data for (c)(5)(i)(3), the entry is made as a permit-required confined space entry;

(E) The determinations and supporting data required by (c)(5)(i)(A), (B) and (C) are documented by the employer and made available to each employee who enters the confined space; and

(F) Entry into the permit space is performed in accordance with (c)(5)(ii).

(ii) The following requirements apply to entry into permit spaces that meet the conditions set forth in (c)(5)(i):

(A) Any conditions that make it unsafe to remove an entrance cover shall be eliminated before removing the cover.

(B) When entrance covers are removed, the opening shall be promptly guarded by a railing, temporary cover or other temporary barrier that will prevent an accidental fall through the opening and that will protect each employee working in the space from foreign objects entering the space.

(C) Before an employee enters the space, the internal atmosphere shall be tested, with a calibrated, direct-reading instrument, for the following conditions in the order given:

(1) Oxygen content

(2) Flammable gases and vapors

(3) Potential toxic air contaminants

(D) There may be no hazardous atmosphere within the space whenever an employee is in the space.

(E) Continuous forced-air ventilation shall be used as follows:

(1) An employee may not enter the space until the forced-air ventilation has eliminated any hazardous atmosphere.

(2) The forced-air ventilation shall be directed as to ventilate the immediate areas where an employee is or will be present within the space and shall continue until all employees have left the space.

(3) The air supply for the forced-air ventilation shall be from a clean source and may not increase the hazards in the space.

(F) The atmosphere within the space shall be periodically tested as necessary to ensure that the forced-air ventilation is preventing the accumulation of a hazardous atmosphere.

(G) If a hazardous atmosphere is detected during entry:

(1) Each employee shall leave the space immediately.

(2) The space shall be evaluated to determine how the hazardous atmosphere developed.

(3) Measures shall be implemented to protect employees from the hazardous atmosphere before an subsequent entry takes place.

(H) The employer shall verify through a written certification, that contains the date, location of the space, and the signature of the person providing the certification, that the space is safe for entry and that the measures required by (c)(5)(ii) have been taken.

- The certification shall be made before entry and made available to each employee entering the space.

(6) When the use or configuration of a non-permit confined space changes, potentially increasing its hazards, reevaluate and reclassify it as a permit-required confined space.

(7) Under certain conditions, a permit-required confined space may be reclassified as a non-permit-required confined space by following these procedures:

(i) If the permit space has no actual or potential atmospheric hazards and if all hazards within the space are eliminated without entry into the space, the permit space may be reclassified as a non-permit confined space for as long as the non-atmospheric hazards remain eliminated.

(ii) If it is necessary to enter the permit space to eliminate hazards, such entry shall be a permit-required confined space entry. If testing and inspection during entry demonstrate that the hazards within the permit space have been eliminated, the permit space may be reclassified as a non-permit confined space for as long as the hazards remain eliminated.

- Control of atmospheric hazards through forced-air ventilation does not constitute elimination of hazards. This is covered in subsection (c)(5).

(iii) The employer shall document the basis for determining that all hazards have been eliminated through a certification that contains:

- The date
- The location of the space
- The signature of the person making the determination

The certification shall be made available to each employee entering the space.

(iv) If hazards arise within a permit space that has been reclassified to a non-permit space under (c)(7), each employee shall exit the space.

- The employer must then reevaluate the space to determine whether it must be reclassified as a permit space.

(8) If an independent contractor must enter permit spaces, prior to entry, the host employer must inform the contractor of:

(i) The permit program requirements

(ii) The permit space hazards

(iii) The safety precautions and procedures that have been implemented by the host employer

(iv) If employees of both the host employer and the contractor will be working in or near the same permit space, the employer must coordinate entry operations with the contractor as required by subsection (d)(11).

(v) Following the entry you must debrief the contractor on the entry operation.

(9) In addition to complying with permit space requirements that apply to all employers, each outside contractor that performs permit space entry operations shall:

(i) Obtain available information on hazards and entry operations from the host employer.

(ii) If employees of both the host employer and the contractor will be working in or near the same permit space, the contractor must coordinate entry operations with the employer.

(iii) Inform the host employer of the permit program followed and the hazards confronted or created during the entry operation.

(d) Permit-Required Confined Space Program

(1) Implement measures to prevent unauthorized entry.

(2) Identify and evaluate hazards before entry.

(3) Develop and implement procedures and practices necessary for safe entry operations that include at least the following:

(i) Specify acceptable entry conditions.

(ii) Isolate the permit space.

(iii) Purging, inerting, flushing or ventilating the permit space as necessary to eliminate or control atmospheric hazards.

(iv) Provide barriers as necessary to protect entrants from external hazards, such as pedestrians or vehicles.

(v) Verifying that conditions are acceptable for the duration of the entry.

(4) At no cost to employees, provide, maintain and ensure proper training and use of the following equipment:

(i) Testing and monitoring equipment needed to comply with subsection (d)(5)

(ii) Ventilating equipment

(iii) Communications equipment necessary for compliance with subsections (h)(3) and (i)(5)

(iv) Personal protective equipment as needed to protect employees

(v) Lighting equipment needed for safe work and emergency exit

(vi) Barriers and shields as required by subsection (d)(3)(iv)

(vii) Equipment, such as ladders, needed for safe ingress and egress

(viii) Rescue and emergency equipment as needed to comply with subsection (d)(9)

(ix) Any other equipment needed for safe entry into and rescue from permit spaces

(5) Evaluate permit space conditions when entry operations are conducted.

(i) Before entry is authorized to begin

- If isolation of the space is unfeasible because the space is large or is part of a continuous system (such as a sewer), pre-entry testing shall be done to the extent feasible and entry conditions shall be continuously monitored in work areas.

(ii) Test or monitor as necessary to determine that acceptable entry conditions are being maintained during the course of the entry.

(iii) When testing for atmospheric hazards, test in the following order:

- Oxygen
- Combustible gases or vapors
- Toxic gases or vapors

Examples are given in Appendices B and E.

(6) Provide at least one attendant outside the permit space for the duration of entry operations.

- Attendants may monitor more than one space as long as they can perform their duties effectively.
- Attendants may be stationed at any location outside the space as long as they can perform their duties effectively.

(7) If multiple spaces are to be monitored by one attendant, the program must provide the means for the attendant to respond to an emergency at one or more of the entry operations without distraction from the attendant's responsibilities at other entry operations.

(8) Designate the persons who have active roles in entry operations, identify their duties and provide training for them as required by subsection (g).

(9) Develop and implement procedures for summoning rescue and emergency services, for rescuing entrants, for providing necessary emergency services to rescued employees, and for preventing unauthorized personnel from attempting a rescue.

(10) Develop and implement a system for preparation, issuance, use and cancellation of entry permits.

(11) Develop and implement procedures to coordinate entry operations when employees of more than one employer are working simultaneously as authorized entrants.

(12) Develop and implement procedures necessary for concluding the entry after entry operations have been completed.

(13) Review and revise the permit space program as necessary to correct deficiencies whenever the employer has reason to believe the employees are not protected. This must happen before any subsequent entries.

- Circumstances requiring this review are:
  - ✓ any unauthorized entry of a permit space,
  - ✓ detection of a hazard not covered by the permit,
  - ✓ detection of a condition not covered by the permit,
  - ✓ the occurrence of an injury or near-miss during an entry,
  - ✓ a change in the use or configuration of a permit space,
  - ✓ employee complaints about the effectiveness of the program.

(14) Review the permit-required confined space program, using the canceled permits, within one year after each entry and revise as necessary, to ensure that employees are protected from permit space hazards.

- Employers may perform one annual review covering all entries performed during a 12-month period.

(e) Permit System

(1) Before entry is authorized, the employer shall document completion of the measures necessary to make the permit space safe by filling out an entry permit.

(2) Before entry begins, the entry permit must be signed by the entry supervisor identified on the permit.

(3) The completed entry permit shall be posted at the entry portal or otherwise be made available for the entrants to see that pre-entry preparations have been completed.

(4) The duration of the permit must not exceed the time required to complete the task listed on the permit.

(5) The entry supervisor shall terminate the entry and cancel the permit when:

(i) The entry operations covered by the permit have been completed.

(ii) A condition not allowed by the permit arises in or near the space.

(6) The employer shall retain the canceled permit for at least one year to facilitate the review of the permit space required by subsection (d)(14).

- Any problems encountered during the entry shall be noted on the permit, so appropriate revisions can be made to the permit space program.

(f) Entry Permit

The entry permit documents that everything required by the permit system is completed before any permit-required confined space entry. The standard gives an example of a confined space entry permit in Appendix D, but any format is acceptable that includes the following information:

(1) The permit space to be entered

(2) The purpose of the entry

(3) Date and duration of the permit

(4) The names of the authorized entrants and a system to track them for the duration of the permit

- A reference to the roster or tracking system used is sufficient.

(5) The name of the attendant(s)

(6) The name of the entry supervisor(s), with the signature or initials of the entry supervisor who originally authorized entry

(7) Lists the hazards of the space

(8) Lists the hazard controls necessary before entry

(9) Lists acceptable entry conditions

(10) Recorded results of initial and periodic tests with the time tested and the names or initials of the testers

(11) Who the rescue and emergency services are and how to call them

(12) Communication procedures for use by attendants and entrants

(13) All equipment needed for the entry

(14) Any other information specific to the entry

(15) Any other permits necessary for the work authorized, such as hot work permits

(g) Training

The employer is responsible for providing the training and certification mandated in this part of the regulation.

(1) The employer must provide the training necessary for employees to safely perform their duties that deal with confined space entry.

(2) Each affected employee shall be trained:

(i) Before assigned to perform confined space entry.

(ii) Before there is a change in assigned duties.

(iii) Whenever there is a change in permit space operations and the employee is not trained in the new procedures.

(iv) Whenever the employer has reason to believe that the permit-required confined space program is not being followed or that the employee is not adequately trained in the procedures.

(3) The specific training required is not listed, but the training must establish employee proficiency in all duties and include new or revised procedures.

(4) The employer must certify that the training has been accomplished. This certification must include the employee's name, the signatures or initials of the trainers and the dates of the training.

(h) Duties of Authorized Entrants

An authorized entrant is an employee who is designated by the employer to enter a permit space. It is the employer's responsibility to ensure that all authorized entrants receive the appropriate training and perform their assigned duties properly as designated under the permit space program.

(1) Know the hazards that may be faced during the entry.

- Employers must ensure that authorized entrants know the hazards of the entry. This includes the mode of exposure to the hazards. Poisons may be inhaled into the lungs, ingested through the mouth or absorbed through the skin.
- Recognize the signs and symptoms of hazard exposure.
- Authorized Entrants must recognize the effects of exposure, particularly the early warning signs and symptoms, so self-rescue can start before help from the outside is needed.
- Understand the consequences of hazard exposure.

(2) Use equipment properly.

- Authorized entrants must be provided with the necessary PPE and other required equipment listed in (d)(4) of the standard and trained to properly use it. It is the responsibility of the employer to assure that it is used.

(3) Communicate with the Attendant.

- Authorized Entrants must maintain contact with the Attendant. This improves their chances of safe exit. Two-way radios or other continuous electronic monitoring equipment, in combination with voice or alarms, is considered effective means of communication. This enables the attendant to monitor the status of the authorized entrants, especially subtle changes in speech or deviation from set communication procedures. The attendant can alert the authorized entrants to evacuate the space if needed.

(4) Alert the attendant of hazards that arise during the entry

(i) Whenever an authorized entrant recognizes any warning signs or symptoms of exposure; or

(ii) Detects a condition prohibited by the entry permit

- The attendant can then alert the other authorized entrants. This allows for an organized evacuation.

(5) Exit the permit space quickly whenever:

(i) An evacuation order is given by the attendant or entry supervisor.

(ii) The entrant recognizes the signs or symptoms of exposure to a dangerous situation.

(iii) The entrant detects a prohibited condition.

(iv) An evacuation alarm is sounded.

Note: There are conflicting interpretations of the regulation by trainers and OSHA alike about members trained as authorized entrants entering permit spaces during the course of an actual rescue without an entry permit. Since it is necessary to take all precautions by performing proper hazard assessment and control before entry, documentation on a written permit should not slow down the rescue and will protect against citations and fines.

(i) Duties of Attendants

The attendant is stationed outside one or more permit spaces to monitor the condition and location of authorized entrants and perform the other duties listed in the employer's permit space program. At least one trained attendant is on duty at all permit entries. Their duties include:

(1) Knowing the hazards that may be faced during the entry.

- Employers must ensure that the attendant knows the hazards of the entry into the permit space. This includes the modes, signs and symptoms and consequences of exposure.

(2) Knowing the behavioral effects of exposure.

- Familiarity with the behavioral effects of hazard exposure gives the attendant the judgment to order the evacuation of entrants if any of the signs or symptoms are detected.

(3) Maintain accurate entrant identification.

- Maintaining an accurate Entrant count ensures that the means used to identify entrants under (f)(4) of the standard can accurately identify those in the permit space. By keeping an accurate count no one can be accidentally left in the permit space, or, in case of emergency, no useless searches are done to look for someone not in the permit space.

(4) Remain outside the permit space.

- Keeping unauthorized individuals out of the permit space, staying alert to hazards, communicating with entrants and summoning rescue services if needed are duties requiring the attendant to remain outside the permit space. Some permit space programs allow the attendant to enter the space for rescue, but only after being replaced by another trained attendant.

(5) Communicate with the entrants.

- Attendants must maintain effective communication with all authorized entrants during the entire entry. This allows the attendant to monitor entrant status and alert them of the need to evacuate if necessary.

(6) Monitor entry activities.

- The Attendant monitors activity both inside the permit space and in the immediate vicinity outside the permit space to determine if it is safe for entrants to remain in the permit space. He orders evacuation whenever he:

  (i) Detects a prohibited condition.

  (ii) Detects behavioral effects of exposure in an entrant.

  (iii) Detects a situation outside the space that could endanger the entrants.

  (iv) Cannot effectively perform his duties.

    - If the Attendant is monitoring more than one entry and has an emergency in one of the them that requires all his attention he must terminate the other entries.

(7) Summon rescue or other emergency services as soon as the attendant determines that authorized entrants may need assistance

- As soon as the Attendant feels that an entrant may need help to exit a permit space, he should immediately begin any non-entry rescue procedures. Making contact with the rescuers he has summoned and briefing them on the hazards and subject information when they arrive are important duties of the attendant.

(8) Prevent unauthorized entry by:

(i) Warning others to stay away from the permit space.

(ii) Advising unauthorized entrants to exit if they have entered the permit space.

(iii) Inform the authorized entrants and the entry supervisor that unauthorized persons have entered the permit space.

(9) Perform non-entry rescues.

- The attendant must have the skills and knowledge to perform non-entry rescues as specified in the permit entry program.

(10) Perform no conflicting duties while acting as the attendant.

- Attendants may perform other duties, such as passing equipment into the space, only if it does not interfere with his primary duties as attendant. If the attendant is monitoring entry into more than one permit space and must focus on a rescue in one of them, he should order evacuation of the other space.

(j) Duties of Entry Supervisors

An entry supervisor is anyone empowered by the employer to authorize or directly supervise entry operations in a permit space.

(1) Know the hazards that may be faced during entry.

- The modes of exposure
- The signs and symptoms of exposure
- The consequences of exposure

(2) Verify that the appropriate entries have been made on the permit, that all tests specified by the permit have been completed and that all procedures and equipment required by the permit are in place before endorsing the permit and allowing entry to begin.

(3) Terminate entry as required by subsection (e)(5).

- When acceptable entry conditions are not present the supervisor must terminate the entry operations and cancel the authorization to enter until acceptable conditions are present.

(4) Verify that rescue services are readily available and the means for summoning them are operable.

- Whether on-site or off-site rescue services are used, they must be available in time to effect a rescue and the means for requesting them should be tested.

(5) Remove unauthorized individuals who enter or try to enter the permit space during entry operations.

- Unauthorized individuals may endanger themselves, authorized entrants or rescue personnel called to rescue them. The entry supervisor has the authority to take appropriate action to deny entry to these individuals.

(6) Determine that entry and work operations remain consistent with the entry permit terms and that acceptable entry conditions are maintained.

- Conditions may change during entry and work in permit spaces. The entry supervisor must determine that acceptable conditions are maintained. If the responsibility for a permit space entry is transferred to another person, the outgoing entry supervisor must determine that acceptable conditions still exist before transferring operational responsibility to the incoming entry supervisor.

(k) Rescue and Emergency Services

(1) An employer who designates rescue and emergency services shall:

(i) Evaluate a prospective rescuer's ability to respond to a rescue summons in a timely manner, considering the hazards identified. What will be considered timely will vary according to the specific hazards involved in each entry. For example §1910.134, Respiratory Protection, requires that employers provide a standby person or persons capable of immediate action to rescue employees(s) wearing respiratory protection while in work areas defined as IDLH atmospheres.

(ii) Evaluate a prospective rescue services ability, in terms of proficiency with rescue-related tasks and equipment, to function appropriately while rescuing entrants from the particular permit space or types of permit spaces identified.

(iii) Select a rescue team or service from those evaluated that:

- Has the capability to reach the subject(s) within a time frame that is appropriate for the permit space hazards identified
- Is equipped for and proficient in performing the needed rescue services

(iv) Inform each rescue team or service of the hazards they may confront when called on to perform rescue at the site.

(v) Provide the rescue team or service with access to all permit spaces from which rescue may be necessary, so that the rescue service can develop appropriate rescue plans and practice rescue operations.

(2) An employer whose employees have been designated to provide permit space rescue and emergency services shall take the following measures:

(i) Provide affected employees with the personal protective equipment (PPE) needed to conduct permit space rescues safely and train affected employees, so they are proficient in the use of that PPE, at no cost to those employed.

(ii) Train affected employees to perform assigned rescue duties. The employer must insure that such employees successfully complete the training required to establish proficiency as an authorized entrant.

(iii) Train affected employees in basic first-aid and cardiopulmonary resuscitation (CPR). The employer shall ensure that at at least one member of the rescue team or service holding a current certification in first-aid and CPR is available.

(iv) Ensure that affected employees practice making permit space rescues at least once every 12 months, by means of simulated rescue operations in which they remove dummies, manikins, or actual persons from the actual permit spaces or from representative permit spaces.

Representative permit spaces shall, with the respect to opening size, configuration, and accessibility, simulate the types of permit spaces from which rescue is to be performed.

(3) To facilitate non-entry rescues, retrieval systems or methods shall be used whenever an entry is made into a permit space, unless the retrieval system would increase the overall risk of the entry or would not contribute to the rescue.

- Retrieval systems must meet the following requirements:

(i) Each entrant shall use a full-body harness, with a retrieval line attached at the center of the Entrant's back near shoulder level, or above the entrant's head, or at another point which the employer can establish presents a profile small enough for the successful removal of the entrant.

  - Wristlets may be used if the employer can demonstrate the full-body harness is unfeasible or creates a greater hazard.

(ii) The other end of the retrieval line shall be attached to a mechanical device or a fixed point outside the permit space in such a manner that rescue can begin as soon as the rescuer becomes aware that rescue is necessary.

  - A mechanical device shall be available to retrieve personnel from vertical type permit spaces more than five feet deep.

(4) If an entrant is exposed to a substance for which a material safety data sheet (MSDS) is required to be kept at the worksite, that MSDS or written information shall be made available to the medical facility treating the exposed entrant.

(l) Employee Participation

(1) Employees shall consult with affected employees and their authorized representatives on the development and implementation of all aspects of the permit space program.

(2) Employers shall make available to affected employees and their authorized representatives all information required to be developed by this section.

Appendices

Appendices A–F provide information and non-mandatory guidelines to assist employers in complying with the appropriate requirements of this section.

A. Confined Space Decision Flow Chart

B. Procedures for Atmospheric Testing

C. Examples of Permit-Required Confined Space Programs

D. Confined Space Pre-Entry Check List

E. Sewer System Entry

F. Rescue Team or Rescue Service Evaluation Criteria

The most current copy of the Federal regulations is available at www.osha.gov.

CHAPTER FOUR

CONFINED SPACE HAZARDS

SCOPE:

This chapter serves as an introduction to the hazards in confined spaces.

TERMINAL LEARNING OBJECTIVE:

At the conclusion of this chapter individuals will be able to identify and describe the common hazards that they might encounter at a confined space rescue incident.

ENABLING OBJECTIVES:

- List the four classes of hazards in a confined space
- List the common reasons for oxygen deficiency in confined spaces
- Describe the common causes of flammable atmospheric conditions in confined spaces
- List the most common toxic chemicals found in confined spaces
- List physical, mechanical and environmental hazards commonly found in confined spaces
- Describe the signs and symptoms of claustrophobia

There are a large number of hazards associated with confined spaces, any one of which can injure or kill entrants and rescuers. These hazards fall into four basic categories: atmospheric hazards, physical/mechanical hazards, environmental hazards and psychological hazards. Before any entry commences, all hazards must be identified and controlled or rendered non-hazardous.

NIOSH published a study titled "Worker Deaths in Confined Spaces," which examined 670 confined space fatalities from 1980 to 1989. The data that we cite in this chapter is from that study and, while a bit dated, is still valid.

Atmospheric Hazards

The NIOSH study found that approximately 65% of employee injuries and deaths occur as a result of hazardous atmospheres. In another more recent study that ran from 1983 to 1993 the number was 80%. Therefore, OSHA standards require that the atmosphere is sampled and monitored before making entry into a confined space for work or rescue purposes.

A **hazardous atmosphere** is an atmosphere that may expose employees to the risk of death, incapacitation, impairment of ability to self-rescue (that is, escape unaided from a permit space), injury or acute illness from one or more of the following causes:

- Flammable gas, vapor or mist in excess of 10% of its Lower Flammable Limit (LFL);
- Airborne combustible dust at a concentration that meets or exceeds its LFL
  - ✓ This concentration may be approximated as a condition in which the dust obscures vision at a distance of five feet or less.
- Atmospheric oxygen concentration below 19.5%, or above 23.5%
- Atmospheric concentration of any substance for which a dose or a permissible exposure limit is published in federal or state requirements which could result in employee exposure in excess of its dose or permissible exposure limit.
  - ✓ An atmospheric concentration of any substance that is not capable of causing death, incapacitation, impairment of ability to self-rescue, injury, or acute illness due to its health effects is not covered by this provision.

- Any other atmospheric condition that is immediately dangerous to life or health

  ✓ For air contaminants for which OSHA has not determined a dose or permissible exposure limit, other sources of information, such as Material Safety Data Sheets (MSDS), published information, internal documents can provide guidance in establishing acceptable atmospheric conditions.

Immediately Dangerous to Life or Health (IDLH) means any condition that poses an immediate or delayed threat to life, that would cause irreversible adverse health effects or that would interfere with an individual's ability to escape unaided from a permit space.

COMMON ATMOSPHERIC HAZARDS IN CONFINED SPACES ARE:

- Deviations in the ambient oxygen level
- Presence of combustible gases, vapors or dusts
- Presence of toxic gases

Ambient Oxygen Level

Air is a mixture of gases. Normal air contains 20.9% oxygen by volume. Deviations from this level, either higher or lower, are a major concern.

Asphyxiating Atmospheres

As we have already stated, asphyxiation is the leading cause of worker deaths in confined spaces. Asphyxiating atmospheres can be termed "simple," which means they contain a gas or gases that are physiologically inert and do not produce any ill effects on the body. While they may not have any toxic characteristics, these atmospheres do not have enough oxygen to sustain human life. They are referred to as oxygen deficient.

Effects of Various Oxygen Levels

Oxygen by Volume	Resulting Condition / Effect on Humans
23.5% and above	Oxygen enriched, extreme fire hazard
20.9%	Oxygen concentration of air
below 19.5%	Oxygen deficient
17%	First signs of hypoxia, deterioration of night vision
16%	Disorientation, impaired judgment and breathing
14%	Faulty judgement, rapid fatigue
8%	Mental failure, fainting
6%	Difficulty breathing, death in minutes

Source: NIOSH

Oxygen Deficiency

In general, this is the primary hazard associated with confined spaces. Breathing oxygen-deficient air causes poor judgment, loss of coordination, fatigue, vomiting, unconsciousness and, ultimately, death. Asphyxiation from insufficient oxygen frequently occurs when subjects, unaware of the problem, reach the point where they cannot save themselves or call for help.

Atmospheres with less than normal amounts of oxygen are common in confined spaces. Oxygen levels below 19.5% by volume are considered hazardous. This deficiency may be due to:

Consumption—which can be caused by:

- Combustion—welding or cutting torches
- Decomposition of organic matter—rotting foods, plant life and fermentation
- Oxidation of metals—rusting

Absorption—which can be caused by:

- The vessel itself or the product stored in the vessel, for example, damp activated carbon

Displacement—which can be caused by:

- Intentional purging with inert gases (nitrogen, carbon dioxide, helium or steam) to remove residual chemicals, gases or vapors
- Unintentional purging by gases that do not support life, for example, engine exhaust

Oxygen Enrichment

When oxygen levels exceed 23.5% by volume, the atmosphere is referred to as oxygen enriched. An oxygen-enriched atmosphere is not an asphyxiation hazard; however, it can be a serious fire hazard. Flammable materials will burn very rapidly in an oxygen-enriched atmosphere. A common cause of oxygen enrichment is leaking oxygen lines or cylinders.

Presence of Combustible Gases

For a fire or explosion to occur, three components must be present simultaneously. The absence of any one will prevent a fire or explosion.

- Fuel (such as a combustible gas)
- Oxygen (in the proper proportions)
- A source of ignition (spark or flame)

The mixture of fuel and oxygen that will ignite is different for each specific combustible gas. This critical point, defined as the **explosive range**, is between the **Lower Explosive Limit (LEL)** and the **Upper Explosive Limit (UEL)**. Concentrations below the LEL, which is the lowest concentration (air/fuel mixture) at which a gas will ignite, are too lean to burn. Concentrations above the UEL, the highest concentration that will ignite, are too rich to burn.

Explosive Range

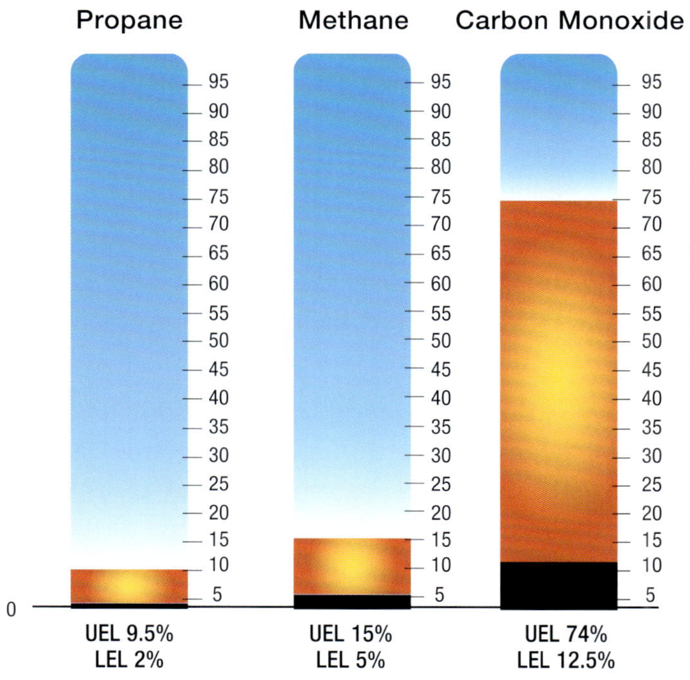

Note: The terms Lower Explosive Limit (LEL) and Lower Flammable Limit (LFL) are used interchangeably, as are Upper Explosive Limit (UEL) and Upper Flammable Limit (UFL.)

Although government standards allow a concentration of flammable contaminants over 10% of their LEL, a concentration of zero is better and is the standard for many industries. However, the atmosphere in a confined space may change as a result of work activities in or around the space. During the rescue operation, any changes in LEL should be investigated, since an upward change in LEL would indicate an increased presence of flammable contaminants in the atmosphere. A concentration above 10% of the LEL is considered hazardous and therefore permit required.

Toxic Atmospheres

Toxic atmospheres in confined spaces can cause serious health problems and even death. Their poisonous physical effects may be immediate, delayed or a combination of both. For example, exposure to low concentrations of carbon disulfide gas causes unnoticed but cumulative brain damage, while high concentrations quickly result in death.

These contaminants usually occur from material previously stored in the space or as a result of the use of coatings, solvents or preservatives. Decomposing organic matter not only displaces and consumes oxygen, it can produce gases, such as methane, carbon monoxide, carbon dioxide and hydrogen sulfide. It is also possible for toxic gases to enter a confined space during an entry because of improper isolation procedures.

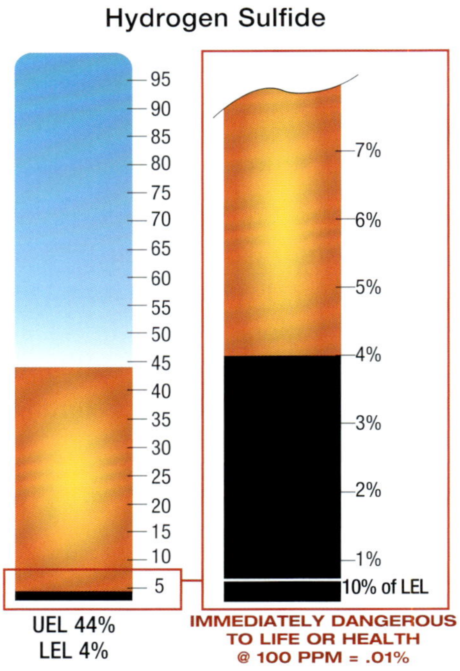

Though some have color and smell, many toxic gases are not detectable by human senses. There are four ways toxics can enter the body: absorption, ingestion, inhalation and injection.

Absorption is possible directly through unprotected skin. Ingestion would be swallowing the material and having it enter through the digestive system. Inhalation, which would be the most common, especially for gases, is by breathing the material and injection would occur from a puncture of the skin due to a sharp object or a fine high pressure stream of fluid.

Permissible Exposure Limit (PEL)

OSHA has developed PELs for most toxic contaminants found in the work place. When an air contaminant for which OSHA has not determined a PEL is present in a confined space, OSHA recommends consulting other sources of information for guidance in establishing acceptable environmental conditions for entry. Some appropriate sources include:

- Safety guidelines issued by NIOSH
- Recommended standards published by the National Fire Protection Association (NFPA)
- Threshold Limit Values (TLV®) developed by the American Conference of Governmental Industrial Hygienists (ACGIH)
- Manufacturer's Material Safety Data Sheets (MSDS)
- Other published information

The five most common gases found in confined spaces are methane, carbon monoxide, carbon dioxide, hydrogen sulfide and sulfur dioxide. They are all life threatening.

Methane (CH_4)—an odorless, colorless, non-toxic combustible gas. It is also known as marsh gas or firedamp. The simplest of hydrocarbons, it is the starting point for many chemical compounds. It is lighter than air with a vapor density of 0.554. It has an explosive range of 5.3% to 15%. It occurs naturally from the decomposition of plant and animal matter and is one of the chief constituents of natural gas.

Carbon Monoxide (CO)—a toxic, odorless, colorless, combustible gas. It is slightly lighter than air with a vapor density of 0.968. It has a wide explosive range of 12.5% to 74%. It is a byproduct of incomplete combustion and is very reactive. It is probably the most widely encountered toxic gas. Carbon monoxide is an intense poison when inhaled. It has an extreme affinity for the hemoglobin in our blood, approximately 200 to 300 times that of oxygen, which it quickly replaces. It is extremely toxic even from the small amounts in the exhausts of internal combustion engines. Signs and symptoms of carbon monoxide poisoning include throbbing headache, nausea, shortness of breath, irritability, confusion, loss of judgment, increased heart and respiratory rate, lethargy and stupor, seizures and changes in skin color form pale to cyanotic to cherry red. The federal OSHA PEL for CO is 50 ppm (parts per million), the Cal/OSHA PEL is 25 ppm and it is IDLH is 1,200 ppm.

Effects of Carbon Monoxide Exposure

All values are approximate. The effects can vary depending on the individual's health and the type of physical activity being performed.

CO Level in ppm*	Resulting Condition / Effect on Humans
0-20 ppm	Grade D breathing air (legal for compressed or airline breathing air)
25 ppm	ACGIH TLV
50 ppm	OSHA PEL
100 ppm	Headache and discomfort after 3 hours
200 ppm	Ceiling level—5 minutes allowable exposure in 4 hours Headache in 2-3 hours
400 ppm	Headache and nausea in 1-2 hours Occipital headache in 2.5-3.5 hours
800 ppm	Headache, dizziness, nausea in 45 minutes Collapse and possible death in 2 hours
1,200 ppm	IDLH level
1,600 ppm	Headache, dizziness, nausea in 20 minutes Collapse and possible death in 1.5-2 hours
3,200 ppm	Headache and dizziness in 5-10 minutes Unconsciousness and death in 30 minutes
6,400 ppm	Headache and dizziness in 1-2 minutes Unconsciousness and death in 10-15 minutes
12,800 ppm	Immediate unconsciousness Death in 1-3 minutes
125,000 ppm	LEL
740,000 ppm	UEL

*ppm—parts per million

Carbon Dioxide (CO_2)—a colorless, odorless, non-combustible gas. It is heavier than air with a vapor density of 1.527. It is non-toxic but will displace oxygen to become an inhalation hazard as an asphyxiant. It is used for fire suppression, inerting atmospheres, carbonated beverages and many industrial processes. In solid form (dry ice) it is used for refrigeration. It is a by-product of distillation and natural gas production. Signs and symptoms of exposure include headaches, dizziness, restlessness and increased respirations, heart rate and acid levels in the blood. The PEL is 5,000 ppm and it is IDLH at 40,000 ppm.

Hydrogen Sulfide (H_2S)—a toxic, colorless, combustible gas that has the odor of rotten eggs. It has an explosive range from 4.5-45.5% and a vapor density of 1.19, making it slightly heavier than air. It is formed by the decomposition of organic plant and animal life by bacteria. H_2S poisons a person by collecting in the blood stream and then paralyzing the area of the brain that controls breathing. As a result, the lungs are unable to function and the individual is asphyxiated. H_2S is found in oil and gas production, refining, sewers, pulp mills and a variety of industries. In low concentrations, H_2S is easily detected by a strong "rotten egg" odor. High concentrations can paralyze the sense of smell. Even low concentrations can effect the olfactory nerves over time, so relying on smell is very dangerous and misleading. In liquid form, H_2S can cause freezing burns. Signs and symptoms of exposure include headache, loss of appetite, dizziness, muscle fatigue and cramps, low blood pressure, loss of consciousness and eventuality respiratory paralysis and death. The PEL is 10 ppm and it is IDLH is 100 ppm. The new ACGIH recommendation for airborne H_2S exposure are a TWA of 1 ppm and an STEL of 5 ppm.

Effects of Various H_2S levels

All values are approximate. The effects can vary depending on the individual's health and the type of physical activity being performed.

H_2S Level in ppm*	Resulting Condition/Effect on Humans
0.13	Minimal perceptible odor
1.00	TWA (ACGIH)
4.60	Easily detectable, moderate odor
5.00	STEL (ACGIH)
10.0	Beginning eye irritation. Permissible exposure limit, 8 hours (OSHA)
27.0	Strong, unpleasant odor, but not intolerable
100	IDLH, coughing, eye irritation, loss of sense of smell after 2 to 5 minutes
200-300	Marked conjunctivitis (eye inflammation) and respiratory tract irritation after 1 hour of exposure
500-700	Loss of consciousness and possible death in 30 minutes to 1 hour
700-1000	Rapid unconsciousness, cessation (stopping or pausing) of respiration, and death
1000-2000	Unconsciousness at once, with early cessation of respiration and death in a few minutes. Death may occur even if individual is removed to fresh air immediately
40,000	LEL
440,000	UEL

*ppm—parts per million

Sulfur Dioxide (SO_2)—a colorless, non-flammable gas with a characteristic irritating, pungent odor. It has a vapor density of 2.264. It is used as a refrigerant, as a preservative, in bleaching and for making other chemicals. The gas is toxic and the suffocating odor makes leaks easily detectable. It is soluble in water. It is an irritant and corrosive to the eyes, nose, throat and skin. It can cause reflex bronchioconstriction, skin burns and edema of the lungs and glottis. Exposure to concentrations of 1 to10 ppm causes respiratory and pulse rate increases and a decrease in the depth of respirations. The PEL is 2 ppm and it is IDLH is 100 ppm.

Characteristics of Selected Chemicals

Gas		Vapor Density	PEL (ppm)	IDLH (ppm)	Explosive Range
Acetylene	C_2H_2	0.91	2500	NA	2.5-100%
Ammonia	NH_3	0.6	25	300	15-28%
Benzene	C_6H_6	2.7	1	500	1.2-7.8%
Carbon Dioxide	CO_2	1.53	5,000	40,000	Non Flammable
Carbon Monoxide	CO	0.97	25	1,200	12.5-74%
Chlorine	Cl_2	2.47	0.5	10	Non Flammable
Ethylene	$CH_2=CH_2$	1	SA	SA	2.75-28.6%
Helium	He	.1368	SA	SA	Non Flammable
Hydrogen	H_2	.0695	NA	NA	4-75%
Hydrogen Cyanide	HCN	0.94	10	50	5.6-40%
Hydrogen Sulfide	H_2S	1.19	10(C)	100	4-44%
Methane	CH_4	.554	SA	SA	5-15%
Natural Gas	-	.55	SA	SA	5-15%
Nitrogen	N_2	.965	SA	SA	Non Flammable
Nitrogen Dioxide	NO_2	1.6	5(C)	20	Non Flammable
Sulfur Dioxide	SO_2	2.264	2	100	Non Flammable
Toluene	$C_6H_5CH_3$	3.2	100	500	1.1-7.1%
Xylene	$C_6H_4(CH_3)_2$	3.66	100	900	1-7%

C Ceiling
IDLH Immediate Danger to Life or Health
PEL Permissible Exposure Limit
ppm Parts Per Million (10,000 ppm = 1%)
SA Simple Asphyxiant
NA Not Available
VAPOR DENSITY Indicates whether the gas is lighter or heavier than air, which has a vapor density of 1.00

Multiple Atmospheric Hazards

The presence of toxic hazards in the air may be entirely independent of oxygen concentrations and flammable contaminants. Many atmospheric hazards, even when reduced below flammable levels, may still be toxic. Others, when reduced below toxic levels, may still be flammable.

Some substances combined with air pose entirely different atmospheric hazards at different concentration levels. For example, methane is not toxic below a concentration in air of 5%, is explosive between 5% and 15% and is asphyxiating above 15%.

Atmospheric Explosive Classifications

Hazardous atmospheres are characterized by the type and form of explosive materials present. The National Electrical Code (NFPA 70) divides the atmospheres into classes, groups and divisions. This is important to the rescuer because they typically use tools, monitors, ventilation, communications equipment and lighting in or near the space that could trigger an explosion if it were not of the proper type. By knowing the ratings of the equipment and matching that to the hazard in the space such an event can be prevented. Not all confined space rescue equipment is rated and it might not be necessary. A non-rated fan set up outside of the space and supplying positive pressure ventilation would not be a concern, but if the fan was inside of the space it could be. Due to the expense of testing the equipment, it might not be rated for every possible atmosphere or hazard.

Two terms used referencing the classification of equipment are "intrinsically safe" and "explosion proof." Intrinsically safe equipment is defined as "equipment and wiring which is incapable of releasing sufficient electrical or thermal energy under normal or abnormal conditions to cause ignition of a specific hazardous atmospheric mixture in its most easily ignited concentration." This is achieved by limiting the amount of power available to the electrical equipment in the hazardous area to a level below that which will ignite the gases. Explosion-proof equipment is sealed to prevent the hazardous atmosphere from reaching an ignition source within the equipment and is manufactured to be strong enough to contain an explosion, if one should occur, within the case of the device.

Class—Indicates the type of atmospheric hazard.

Class I— Locations in which flammable gases or vapors are or may be present in the air in quantities sufficient to produce explosive or ignitable mixtures.

Class II—Locations which are hazardous because of the presence of combustible dust.

Class III— Locations which are hazardous because of the presence of easily ignitable fibers or flyings but in which such fibers or flyings are not likely to be in suspension in the air in quantities sufficient to produce ignitable mixtures.

Group—Classes are further divided into groups based on similar flammable characteristics. Groups A-D fall under Class I, Groups E-G fall under Class II.

Group A— Atmospheres containing acetylene.

Group B— Atmospheres containing hydrogen, fuel and combustible process gases containing more than 30% hydrogen by volume or gases, vapors or equivalent hazard, such as butadiene, ethylene oxide, propylene oxide and acrolein.

Group C—Atmospheres containing ethyl ether, ethylene or gases or vapors of equivalent hazard.

Group D—Atmospheres containing acetone, ammonia, benzene, butane, cyclopropane, ethanol, gasoline, hexane, methanol, methane, natural gas, naphtha, propane or gases or vapors of equivalent hazard.

Group E—Atmospheres containing combustible metal dusts, including aluminum, magnesium and their commercial alloys or other combustible dusts whose particle size, abrasiveness and conductivity present similar hazards in the use of electrical equipment.

Group F—Atmospheres containing combustible carbonaceous dusts, including carbon black, charcoal, coal or dusts that have been sensitized by other materials so that they present an explosion hazard.

Group G—Atmospheres containing combustible dusts not included in Group E or F, including flour, grain, wood, plastic and chemicals.

Division — Indicates the occurrence of release of the explosive material.

Division 1—The hazardous material is more or less freely present in the air in conjunction with manufacture.

Division 2—The hazardous material is confined in containers, and explosive mixtures with air will occur only in case of accident or through failure of ventilation systems to operate properly.

PHYSICAL/MECHANICAL HAZARDS

Engulfment

Engulfment means the surrounding and capture of a person by a liquid or finely divided (flowable) solid substance. According to the NIOSH study, engulfment was the number two cause of death in confined spaces. The study shows as many as 227 deaths attributed to engulfment between 1980-1989. Engulfment and suffocation are hazards associated with storage bins, silos and hoppers where grain, sand, gravel, coal, flour, cement, sawdust or other loose materials are stored, handled or transferred. The behavior of these materials is unpredictable, and entrapment and burial can occur in a matter of seconds.

A condition known as bridging can create additional hazardous situations. Bridging occurs when grain or other loose material clings to the sides of a container or vessel that is being unloaded from below, allowing a hollow space to be created. The bridge of material over the space may collapse without warning, entrapping workers who are standing below or on top of the bridge and who are unaware the surface is unstable. Bridging can occur in storage bins, silos and hoppers. The diameter of the storage vessel and the moisture content and temperature of the stored materials are factors that contribute to bridging. To prevent such an occurrence, access to the top of the material should be limited and safety/retrieval lines are required on all

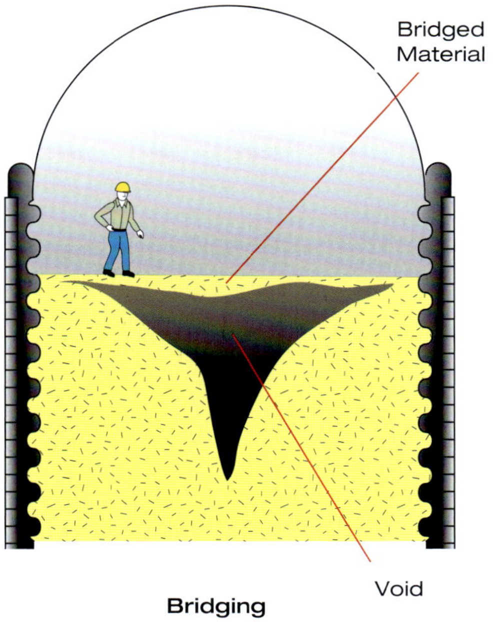

Bridging

personnel within the space. Engulfment accidents can also occur during the accidental release of product into a container or from a bottom dump container when employees are working below.

Engulfment hazards either block the respiratory system or compress the upper body to the point where death occurs due to mechanical asphyxiation (suffocation) or crushing.

While the NIOSH study did not include agricultural incidents, those are quite common and should be recognized by rescuers, especially in rural regions. The most common bridging fatalities occurred with grain followed by sand, gravel, cement and other types of building materials. We illustrate bridging with a silo type of container, but bridging can occur in large open piles as well. There have been rare cases where bulldozers have fallen through bridged coal at power plants and a backhoe through wood chips at a paper mill. In those cases, the operators were protected by the cab of the equipment but access to them was still difficult and engulfment was a constant hazard to the rescuers.

Mechanical Hazards

There are many types of mechanical hazards that exist in confined spaces. Grinding equipment, weirs, agitators, mixers, augers, drive shafts and gears are a few examples of moving parts that can maim or crush entrants. Rams can be powered by hydraulic or pneumatic means. Presses rely upon kinetic energy or gravity. Electrical power is usually in the area for lighting, motors or as a heat source. Uneven or wet surfaces may pose slip, trip or fall hazards. Wet surfaces also increase the hazard of electrocution because they can provide a grounding path in the event of accidental contact with energized circuits.

Environmental Hazards

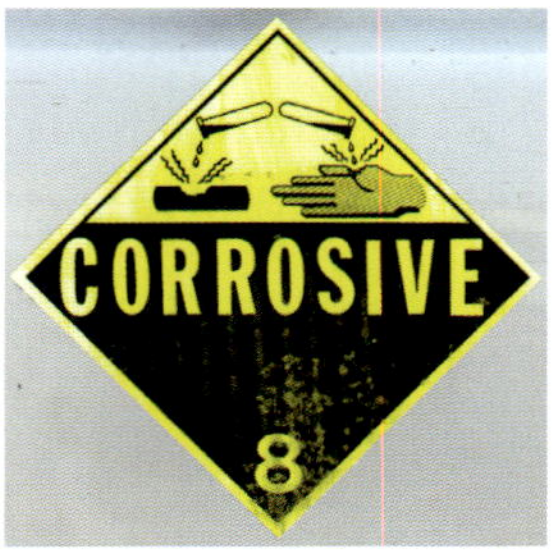

Corrosive Hazards

Corrosive chemicals, such as acids and cleaning solutions, can pose yet another confined space hazard. Contact between these substances and the skin, mucous membranes or eyes can cause serious irritation or burns. Fumes released by these materials can also irritate the respiratory system and cause gastrointestinal distress.

Temperature Hazards

All confined spaces have the potential for elevated temperatures, especially during the hot summer months. This presents the hazard of heat stress for the entrants, a quickly debilitating factor in work efficiency and rescuer safety. Entrants should undergo medical monitoring before and after entry to insure their safety. Normal precautions for operations at elevated temperatures, such as limiting work periods and fluid replacement, are appropriate. Many industrial processes require the use of high temperature equipment. Entrants should be aware and protect themselves from the potential for burns from steam or hydraulic lines, burners, ovens, etc. If a space has been steam-cleaned, it should be allowed to cool before entry is made.

Biological Hazards

Biological hazards, such as molds, mildews and spores frequently found in dark, damp spaces, can irritate the respiratory system. Bacteria and viruses, found in processes such as sewage treatment, can also threaten the body with a variety of diseases. Insects and animals, such as rats, skunks, snakes and spiders, can bite an entrant who places their hand or foot in the wrong place and inject venom or infect the wound with rabies or other diseases. In addition, bird, bat and animal feces may present serious health hazards to humans, and decaying biological materials can cause oxygen deficiency.

Radiation Hazards

Radiation hazards can exist from the accidental release of chemicals due to an event, such as an explosion or earthquake. In certain industrial sites and hospitals, radioisotopes are used but these are rarely encountered. These should be identified during preplanning.

Psychological Hazards

Claustrophobia

Claustrophobia is considered a psychological hazard because of the potential debilitating effects it can have on entrants to a space.

From the French word *claustrum*, meaning a confined area, claustrophobia is an abnormal fear of a closed or confined space. Although a phobia is defined as an abnormal fear, claustrophobia may effect all entry personnel at one time or the other. Its onset is as likely with rescuers with extensive confined space entry experience as it is with those with little experience. For this reason, rescuers need to understand this condition to enable them to control their response.

The most important thing about a phobia is understanding how it occurs and how to recognize its signs and symptoms. The phobic reaction is an expanding chain reaction, complicated by the fact that failure to recognize these signs and symptoms could result in total loss of control.

Involuntary stress reaction is normal for all people and is manifested by the fight-or-flight syndrome, or in technical terms, the body's sympathetic response. Objectively this may be recognized by:

- Elevated pulse
- Increased respiration
- Sweating
- Clammy, cold palms

This initial stress response automatically creates panic, which if not recognized, leads to the chain reaction. This may be subjectively recognized by:

- The feeling you have no air!
- The feeling that the walls are closing in on you!
- The feeling that you are lost!

These stress and anxiety feelings will cycle up and down over the progression of this reaction. By confronting and controlling these symptoms and conditions, the rescuer should be able to continue toward completing the rescue task.

The actual hazard occurs when they act on those fears and, for example, remove their breathing apparatus in a hazardous atmosphere, bolt from the space or just become so debilitated that they are no longer able to perform the rescue and become part of the problem.

Fatigue

Confined space rescues are tiring. The protective equipment and clothing required is heavy and cumbersome. Maneuvering through small openings and in tight spaces is difficult. The level of mental performance must be very high. All of these factors contribute to fatigue. Entrants and supervisors must be aware of the conditions and their effect on safety and performance.

High Noise Levels

Noise within confined spaces can be amplified because of the design and acoustic properties of the space. Excessive noise is not only harmful to the worker's hearing, but can also affect communication and cause warnings to go unheard. Excessive noise can also contribute to the mental stress that contributes to worker fatigue.

Hazard Identification

Rescue personnel must be able to identify all potential hazards connected with the confined space. Potential hazards are everywhere, but some industries and spaces should be examined with extra caution before entry: locations with mechanized assembly equipment; spaces with electrical, hydraulic or pneumatic energy supplied to it or within it; grain or other particulate matter production or processing; sites with top-load storage or processing applications or bottom-dump applications.

CHAPTER FIVE

ATMOSPHERIC MONITORING

SCOPE:

This chapter serves as an introduction to atmospheric monitoring.

TERMINAL LEARNING OBJECTIVE:

At the conclusion of this chapter, individuals will describe different atmospheric monitoring instruments, selection and use of instruments, monitoring strategies and locations in/around confined space rescue incidents.

ENABLING OBJECTIVES:

- Describe the purpose of performing atmospheric monitoring and what may be determined based upon the instrument readings
- Describe three different types of atmospheric monitoring and what may be determined based upon the instrument readings
- Describe three different types of atmospheric monitoring instruments and the reasons for monitoring for specific hazards in sequence
- Describe four different atmospheric monitoring strategies
- Describe stratification of atmospheres within a confined space and the importance of monitoring at different levels within the space

We learned in the previous chapter that airborne contaminants are the most common hazard encountered by workers in confined spaces. In rescue situations things are no different. Specific information about the types and levels of atmospheric hazards must be readily available to rescue personnel. Air monitoring is an essential component of confined space rescue operations. Reliable measurements of airborne containments are useful for:

- Determining the survival profile of subject
- Selecting proper PPE
- Delineating areas where protection is needed
- Assessing the potential health effects of exposure
- Determining the need for specific medical monitoring

Gas-Monitoring Instruments

Gas Monitor

Modern environmental monitoring instruments provide read out information in real time, that is, at the time the monitoring is being performed. They are available from single gas models to as many as five separate gases being monitored at the same time. Oxygen and flammability are included as standard on multi-gas monitors with the remaining space filled with sensors to target general or industry -specific hazards. Carbon monoxide and hydrogen sulfide are common for emergency responders, but sulfur dioxide, carbon dioxide, chlorine, ammonia, nitrogen dioxide or dozens of other gas specific sensors may be selected.

To be useful in the field, air-monitoring instruments must be:

Portable—This is a prime consideration for any field instrument. Transportation and shock, together with unintentional abuse, greatly shortens the useful life of an instrument. To reduce this trauma, instruments should be selected that have reinforced shells or frames, shock-mounted electronic packages or padded containers for shipping.

A portable unit should possess ease in mobility, the ability to withstand the rigors of use, quick assembly and short checkout and calibration time. Also, some instruments, though portable, are still too large to be useful in confined spaces, so size is important.

Able to generate reliable and useful results—Response time, the interval between an instrument sensing a contaminant and generating data, is important to producing reliable and useful results in the field. Response time depends on tests to be performed, dead time between sample periods (the time for analysis, data generation and data display) and the sensitivity of the instrument. Response time establishes the pace of the overall survey and the individual tests.

Another consideration is that the instrument must provide results that are immediately useful. Instruments should be direct reading and easy to use with little or no need to interpolate, integrate or compile large amounts of data.

Sensitive and have selective results—A good field instrument must have the ability to sample and analyze very low contaminant levels and to distinguish among contaminants exhibiting similar characteristics.

Sensitivity defines the lowest concentration of a contaminant an instrument can accurately and repeatedly analyze. In the strictest sense, it is a function of the detecting ability of the instrument. The operating range establishes the upper and lower use limits of the instruments.

Selectivity establishes what contaminants will give a response on the instrument. Selectivity also determines which, if any, interferences may produce a similar response. Selectivity and sensitivity must be reviewed and interpreted together. Many devices have a high selectivity but widely varying sensitivities for a given family of chemicals, for example, aromatics, aliphatics and amines.

Amplification, often used synonymously (and incorrectly) with sensitivity, deals with an electronic amplifier's ability to increase very small electrical signals emanating from the detector. This capacity may be fixed or variable. However, changing the amplification of the detector does not change its sensitivity.

For optimum field usefulness, an instrument should possess high sensitivity, strong selectivity and the ability to vary the amplification of detector signals.

Intrinsically safe—A field instrument must not introduce a dangerous ignition source into a potentially flammable atmosphere.

All direct-reading instruments have inherent constraints in their ability to detect hazards:

- They usually detect and/or measure specific classes of chemicals
- Generally, they are not designed to measure and/or detect airborne contaminants below 1 ppm
- Many of the direct-reading instruments that have been designed to detect one particular substance also detect other substances (interferents) and, consequently, may give false readings

It is imperative that direct-reading instruments be operated and have their data interpreted by qualified individuals who are thoroughly familiar with the particular device's operating principles and limitations and who have also obtained and are familiar with the device's latest operating instructions and calibration information. At the scene of a confined space rescue, instrument readings should be interpreted conservatively. The following guidelines may facilitate accurate recording and interpretation:

- Calibrate instruments on a regular basis according to the manufacturer's instructions. Time, humidity, temperature and elevation all can affect the accuracy of readings. We advocate that when time permits and qualified personnel are available that instruments be recalibrated before use by rescue personnel.
- Remember that the instrument readings have limited value where contaminants are unknown. Conduct additional monitoring at any location where positive response occurs.
- A reading of zero should be reported as "no instrument response" rather than "clean" because qualities of chemicals may be present that are not detectable by the instruments.
- When monitoring unknown atmospheres, the air-monitoring survey should be repeated with several different detection systems to maximize the number of chemicals detected.
- Use multiple monitors whenever possible. This allows a confirmation of the readings, either positive or negative, for greater safety.

Monitor Start-up Procedure

Each brand of instrument will be slightly different in its start-up procedures. **Know your instrument.** This is a generalization of the process.

1) Turn the monitor on in a clean atmosphere. Be sure to choose a well-ventilated area away from vehicles, generators and other exhaust sources.

2) Let it go through its self-diagnosis process. This may take a minute or two while it checks the battery level, the various sensors, the pump, possibly clears the peak readings and memory (or asks you if it should). Some brands give you a progress report and countdown timer of the process on the screen.

3) Install the sample hose, place the unit in a pump adapter and/or turn on the pump as required. Test the low flow alarm function by putting your finger over the end of the hose or sampling tube.

4) Verify the instrument is operating properly by performing a bump test using alarm check (bump) or calibration gas. Match your monitor readings with the contents of the cylinder listed on the label. Be sure that you perform this test in a well-ventilated area and clear the peak values at the conclusion of the test. Remember the bump test is to insure that the alarms are working and that the monitor reads within an acceptable range. It is not a substitute for a full calibration.

Monitors for Oxygen

Monitoring to find the oxygen content in the space is the first priority. In reality, with modern multi-gas monitors this happens automatically. They check for all of their target gases simultaneously. It is critical that the environment contain an adequate level of oxygen before accurate readings can be obtained.

The oxygen section of multi-gas monitors is used to detect the percentage of oxygen in the atmosphere. Most oxygen-sensing devices are calibrated to indicate concentrations between 0% and 25%. This is the most useful range for response personnel since decisions involving PPE and combustible indicators fall within this range.

Oxygen detection occurs in an electrochemical cell consisting of two electrodes in an alkaline electrolyte covered by a membrane. Oxygen enters through the membrane, reacts with the electrolyte and sends a current and voltage to the detector terminals. This voltage is interpreted as percentage of oxygen in the air being monitored.

ADVANTAGES

- Quick response time
- Portable (remote cables and hoses available)
- Single monitors or combination monitors available

DISADVANTAGES

- Must be calibrated at the temperature, humidity and elevation of use
- Oxygen meters are adversely affected by certain gases and vapors. In particular, carbon dioxide can permanently affect meter response. As a general rule, oxygen meters can be used in atmospheres greater than .5% CO_2 only with frequent replacement of the oxygen detector cell.

Monitors for Flammable or Explosive Atmospheres

The second monitoring priority is for flammable or explosive atmospheres. It is essential to be able to recognize flammable or explosive conditions and equally as important to be able to anticipate the potential for such an atmosphere. This can be accomplished by using an instrument capable of measuring for these atmospheres. There are three different scales used on various meters to display the level of combustible vapors in the atmosphere:

- ppm
- % gas
- % LEL

The most common is the % LEL scale. The explosive range of a gas is between the LEL and the UEL.

- Below LEL—too lean to burn
- Above UEL—too rich to burn

The % LEL scale is 0 to 100% of the LEL, not 0 to 100% of the explosive vapor or gas in the air. For example, if the reading is 50% LEL, then it indicates that 50% of the flammable vapor necessary to support combustion is present in the tested air. If the LEL of the gas is 2%, then the instrument is indicating that there is one-half (50%) of 2% (that is, 1%) gas present. If the concentration increases, it will get closer to the LEL. While some instruments also indicate when the concentration of combustible gas or vapor in air exceeds the UEL, not all do. When using instruments that do not read above the UEL, watch the meter while the air sample is diffused on the sensor. The LEL may rise until it exceeds the UEL and then drop back down to 0%. This can indicate that the atmosphere is too rich to burn and the possibility that supply ventilation will bring it down to the explosive range.

Explosive meters are able to detect gases or vapors by either allowing air to diffuse into the sensor or by drawing a sample of air through a rigid probe or hose. A hand-operated or battery-powered pump may be used to draw the sample. There are several types of instruments used to detect these atmospheres.

Combustible Gas Indicators

Combustible Gas Indicators (CGI) diffuse the sample atmosphere through a coarse metal filter, where it then comes in contact with two heated platinum filaments inside the sensor. Both filaments are heated to the same temperature and therefore have the same resistance. One filament is plain and the other is coated with a catalyst. The combustible gases burn on the catalyst-coated filament but no combustion occurs on the uncoated filament. The combustion causes the filament with the catalyst coating to increase in temperature, resulting in an increase in resistance. This change in resistance causes an imbalance in the monitor's resistor circuit called a Wheatstone bridge. The change in resistance is translated into a CGI meter reading.

All CGI readings are relative to the calibrated gas. The readings correspond to the relative increase in resistance produced by the calibrated gas when it burns on the catalytic filament. When measuring another gas or vapor, the instrument still responds to the increased temperature of the filament. However, some vapors and gases produce more heat when burned. These hot-burning gases cause the catalytic filament to become hotter at lower concentrations than the calibrated gas. Conversely, some gases burn cooler than the calibrated gas, and a higher concentration of such a gas is needed to cause the same increase in filament temperature.

Different gases require a response curve to the calibration gas. If the calibration gas is pentane, then a relative response curve would be used to give actual concentrations of the gas being metered. Methane burns hotter than pentane, so the meter reading is at 100% LEL when the actual concentration is less than 70% LEL. Xylene burns cooler, so the meter reads less than 50% LEL when a 100% LEL condition exists.

Flammable Gas Conversion Chart

Calibration Gas

Gas Being Sampled	Acetone	Acetylene	Butane	Hexane	Hydrogen	Methane	Pentane	Propane
Acetone	1.0	1.3	1.0	0.7	1.7	1.7	0.9	1.1
Acetylene	0.8	1.0	0.7	0.6	1.3	1.3	0.7	0.8
Benzene	1.1	1.5	1.1	0.8	1.9	1.9	1.0	1.2
Ethane	0.8	1.0	0.8	0.6	1.3	1.3	0.7	0.8
Ethanol	0.9	1.1	08	0.6	1.5	1.5	0.8	0.9
Ethylene	0.8	1.1	0.8	0.6	1.4	1.3	0.7	0.9
Hexane	1.4	1.8	1.3	1.0	2.4	2.3	1.2	1.4
Hydrogen	0.6	0.8	0.6	0.4	1.0	1.0	0.5	0.6
Isopropanol	1.2	1.5	1.1	0.9	2.0	1.9	1.0	1.2
Methane	0.6	0.8	0.6	0.4	1.0	1.0	0.5	0.6
Methanol	0.6	0.8	0.6	0.5	1.1	1.1	0.6	0.7
Pentane	1.2	1.5	1.1	0.9	2.0	1.9	1.0	1.2
Propane	1.0	1.2	0.9	0.7	1.6	1.6	0.8	1.0
Styrene	1.3	1.7	1.3	1.0	2.2	2.2	1.1	1.4
Toluene	1.3	1.6	1.2	0.9	2.1	2.1	1.1	1.3
Xylene	1.5	2.0	1.5	1.1	2.6	2.5	1.3	1.6
JP-4							1.2	
JP-5							0.9	
JP-8							1.5	

Example: The instrument has been calibrated on methane and is now reading 10% LEL in a pentane atmosphere. To find actual % LEL pentane, multiply by the number found at the intersection of the methane column (calibration gas) and the pentane (gas being sampled)... in this case, 1.9. Therefore, the actual LEL pentane is 19% (10 x 1.9).

Source: Industrial Scientific

It is also important to emphasize that both an oxygen meter and a CGI are necessary for assessing an unknown environment because it is possible to detect a high % LEL concentration without a significant change in oxygen concentration. Also, oxygen is required for proper functioning of any CGI since oxygen is necessary for the combustion of the gas or vapor. Most manufacturers indicate that the instrument will not give an accurate reading at less than 10% oxygen. Oxygen-enriched atmosphere will enhance the catalytic process and will result in high readings that are false.

The catalytic filament is vulnerable to contaminants, such as sulfur compounds, moisture and high-voltage power lines.

As you learn more about different monitoring equipment, you will see that other meters detect combustible gases, but are referred to as other than CGI. In the instrument world, only those with a Wheatstone bridge are considered CGIs.

Broad Range Instruments

Another instrument sensing technology uses a Metal-Oxide Semi-conductor (MOS). MOS sensors have an advantage over catalytic filaments in that they operate at lower oxygen levels. MOS sensors are also sensitive to a wider range of flammable materials and detect them at lower levels. This makes them the preferred instrument to use when actual or potential contaminants have not been identified. They are also less prone to being poisoned by lead- and silicon-containing materials.

ADVANTAGES OF BOTH CGI AND MOS INSTRUMENTS

- Fast response
- Simple operation
- Audible and visual alarms available
- Extension hoses and adapters available
- Portable
- Choice of active or passive sampling

When combined with an oxygen monitor:

- Measures 0% to 25% oxygen
- Incorrect readings are more readily identified by comparing them to oxygen content readings

DISADVANTAGES OF CGI INSTRUMENTS

- The reaction is temperature dependent. The closer the temperature of the calibrating atmosphere is to that of the sampling atmosphere, the more accurate the measurement.
- Calibrated to one gas. If other gases or vapors are present, readings will be higher or lower than the actual gas concentrations.
- Leaded gasoline vapors, halogens, sulfur or silicon compounds may damage the detector element.
- The instrument will give erroneous readings in atmospheres containing 10% or less oxygen.

Monitoring Start-up Procedure

The start-up procedures for CGI and MOS instruments are the same as described for other gas-monitoring instruments.

Monitors for Toxic Contaminants

The third priority for atmospheric monitoring is to test for toxic substances. There are hundreds of potential toxic substances that can be present in concentrations that are IDLH.

CO and H_2S are two of the dozens of toxic gas sensors available. Those particular chemicals are commonly measured with single gas meters as well as with the multi-gas monitors carried by rescue personnel. Toxic gas sensors are similar to oxygen sensors in that they detect through an electrochemical reaction with the gas. Like oxygen meters they are also subject to interference from other gases or vapors. They are also affected by environmental conditions, such as temperature and barometric pressure. Remember, your life and the life of your team members is at stake. Read and understand the operating instructions for your particular instrument.

Colormetric Detector Tubes

Colormetric tubes are very useful for measuring the concentration of known vapors or gas contaminants in the air. Detector tubes respond to a specific chemical or group of chemicals. The concentration of contaminant is determined by observing the color change in the tube. Tubes are available for immediate sampling or for long-term monitoring.

Colormetric indicator tubes are glass tubes filled with a chemical (reagent). The tips of the glass tube are broken off and the tube is then connected to an aspirating pump. The pump may be a piston-type or bellows-type pump. Many tubes have an arrow to indicate the direction of air flow through the tube.

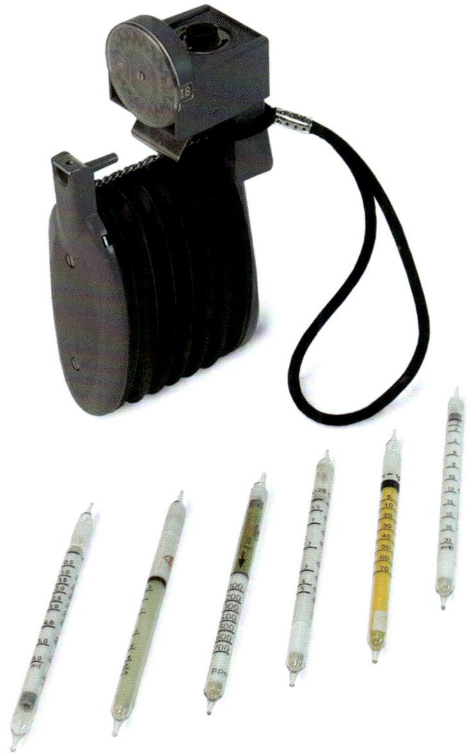

Hand-Sampling Pump and Color Metric Tubes

The vapor or gas contaminant reacts with the reagent, resulting in a color change, or stain, in the tube. Depending on the tube selected, the scale may read in ppm or percent as indicated by the length of the stain. Detector tubes are gas specific, that is, they are calibrated to one material, but they typically respond to many other chemicals that are similar in structure and reactivity. The color of the stain must be appropriate for the gas tested. Off-color stains may indicate the presence of other toxic vapors. Some manufacturers also produce tubes for groups of gases, for example, aromatic hydrocarbons.

Instructions should list potential interferants that can cause inaccurate tube readings. The chemical in the tube can react with air contaminants in addition to the gas or vapor to which it is calibrated. A color change in the tube that is different from that expected may indicate that some other vapor or gas is present. Also interfering gases or vapors may increase or decrease tube response.

Finally, environmental conditions, such as barometric pressure, humidity and temperature, can have a direct effect on the chemical reaction in the tube by increasing or decreasing the tube response. Manufacturers instructions may include temperature and humidity conversion factors, or with some tubes, a filter tube will be used with the test tube.

ADVANTAGES

- Simple to use
- Rapid and convenient
- Relatively quick response
- Over 230 types of tubes available
- Portable
- Inexpensive to use

DISADVANTAGES

- Some tubes will indicate the presence of other gases in the air. This is called an interfering gas or vapor, or interferant.
- Accuracy + or – 25%. The endpoint of the stain is not a definite line.
- Two-year shelf life. Refrigeration helps maintain accuracy.

Photoionization Detectors

Photoionization detectors (PID) are general survey instruments designed to detect organic vapors and gases in the low ppm range. They also detect a small number of inorganic gases, though many of these inorganics are halogens or halogenated compounds. Since these compounds are corrosive, they can damage instruments that are not corrosion resistant. Most PIDs are not resistant.

PIDs are useful for general atmospheric monitoring, characterizing release plumes, screening samples and evaluating relative differences in concentrations from one location to another. They cannot determine the identity of unknowns in the air. They indicate only that there is a high or low concentration of a chemical present.

Flame Ionization Detectors

Flame ionization detectors (FID), also called organic vapor analyzers (OVA), are versatile monitoring instruments. These meters come with ppm readings ranging from 0 to 10,000 ppm. Depending on the model, this instrument can be used for general surveys or as a qualitative instrument that can assist the user in determining the identity of a sample. Use of the FID as a qualitative instrument requires additional training, skill and knowledge.

Application of Detection Devices

In addition to knowing how to use the equipment, it is important to keep in mind that detection equipment will not make decisions. Like all equipment, these detection and monitoring devices each perform specific functions and are limited in the information they can provide. Further, if the equipment is not properly maintained and operated, the information obtained may be incorrect. The operator of the detection equipment must be well trained and aware of the responsibility to ensure that the equipment is used within its limits.

Selection of the proper detection device depends on the suspected hazard. No one meter will detect all hazards; however, some meters detect multiple hazards. For known and unknown atmospheres the first consideration should be given to an explosive atmosphere; but remember, you need to know the amount of oxygen present to know how accurate the combustible meter is reading.

There are factors that directly affect the readings of each instrument. Some of these factors are inherent in the instrument, others will be determined by the environment in which the instrument is used:

- Proper equipment operation
- Instrument calibration and calibration checks
- Equipment detection range
- Device relative response (compared to the calibration gas)
- Response time
- Inherent safety
- Nature of the hazard
- Environmental conditions

If any of the factors below are not addressed, the readings may be improperly evaluated and result in poor decision making:

- Zero out and bump test prior to use
- Use appropriate instrument
- The absence of evidence is not evidence of absence
- Never assume only one hazard is present
- Use one instrument to confirm another
- Establish action levels
- Obtain continuous readings
- Use conservative judgment in interpreting readings

Selection of Detection Devices

To ensure personnel safety, it is recommended that only instruments approved by a National Recognized Test Laboratory (NRTL), such as Factory Mutual (FM) or Underwriters Laboratory (UL), be used in confined spaces and only in atmospheres for which they have been approved. When used in atmospheres with unknown hazards, the monitoring instrument should be rated for use in the most hazardous of locations. For most confined space work this means that devices approved for Class I (vapor and gases), Division 1 (areas of ignitable concentrations), Groups A, B, C and D should be chosen whenever possible.

An additional consideration is that all instruments used in a methane environment should be approved by the Mine Safety and Health Administration (MSHA) as being safe in such atmospheres.

For more information on preventing ignition, see Chapter 6, Hazard Control.

This label on a gas monitor shows details of approval.

Monitoring Strategies

Priorities for air monitoring of confined spaces should be based on the information gathered during the initial site characterization. This information serves as the basis for selecting the appropriate monitoring and PPE to use when conducting site monitoring and confined-space rescue. Depending on site conditions and rescue goals, four categories of monitoring may be necessary: (1) general site monitoring, (2) perimeter monitoring, (3) confined space monitoring and/or (4) personal monitoring. Regulation requires that monitoring be done in a specific order:

- Oxygen
- LEL
- Toxicity

It is necessary to know the oxygen content of the atmosphere to determine the validity of the LEL readings. A lower than normal oxygen reading may indicate the presence of toxic contaminants that you may not have the capability to test for. This will be discussed later in this chapter.

The results of all monitoring in and around the space must be recorded. Be sure to note the location, time results and the name of the person doing the air monitoring. In the OSHA investigation that will follow an accident, OSHA's perspective is "if it isn't documented, it didn't happen."

Immediately Dangerous to Life or Health

Monitoring should be conducted to identify any IDLH conditions. Some IDLH conditions will cause the immediate collapse of personnel subjected to them. Others may produce immediate transient effects that may pass without medical attention, but are followed by sudden, possibly fatal collapse 12 to 72 hours later. Both types of exposure must be considered immediately dangerous to life or health.

General Site Monitoring

General site monitoring consists of monitoring all areas around the confined space where rescue and support personnel may be staging or working. Start upwind of the space and wear PPE for any suspected hazards. Subsequent sampling and/or rescue operations should be based on the results obtained and the potential for an unexpected release of hazardous substances. Extreme caution should be exercised in continuing a site survey when any atmospheric hazards are detected. Monitoring personnel must be aware that conditions can suddenly change from non-hazardous to hazardous.

Perimeter Monitoring

Air monitoring around the perimeter measures contaminant migration from the space and enables the safety officer to determine where PPE is or is no longer required. This is a continuous process and conditions can change due to ventilation of the space, wind, etc.

Confined Space Monitoring

If possible, identify the sources of possible contaminant generation. If ventilation has been established prior to your arrival, shut it down while you monitor the space. Sample the area around the opening of the space, and then just inside the opening, if possible, without opening the space. It may be necessary to use a probe or hose attached to the monitor and drawing air to the sensors with a hand-operated or electric pump. Stick the probe through small holes or cracks in the space. If it is necessary to open the space, be aware that toxic contaminants may immediately move out of the space, so take the proper precautions. Test all possible areas within the space that can be tested while remaining outside. Acutely hazardous concentrations of chemicals and their vapors may remain in a confined space for long periods of time. The recommended practice is to drop the sampling tube to the bottom of the space and then slowly pull it out, taking readings approximately every four feet. Remember to allow for the delay caused by the hose. Horizontal spaces can be tested by attaching the hose to a pole to reach into the space. Remember, if entering a confined space is necessary to conduct the air monitoring within the space, the entry must be conducted as a permit-required entry. The regulation defines entry as any part of the body breaking the plane of the opening.

Personal Monitoring when Entering a Confined Space

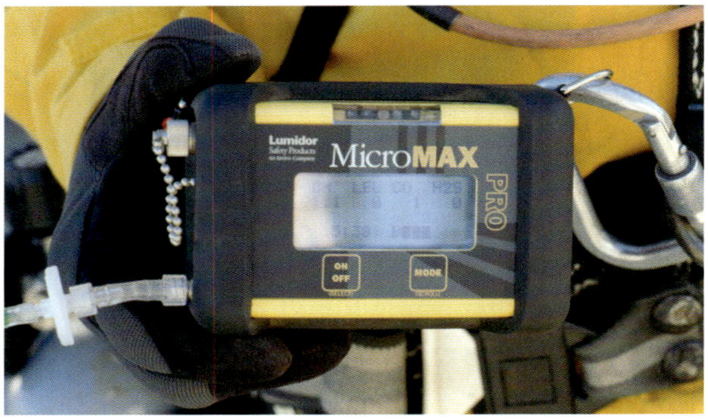

At least one of the rescuers entering the space should also have a monitor with them. There are several reasons for this. The first is that the probability of significant exposure is greatest for those inside the space and closest to the hazard. The configuration of the space might not allow thorough monitoring from the outside. The entrants can monitor for their personal safety, but they can also communicate those readings to the attendant. When the entry team reaches the subject, they can also give a reading at that location which will help determine the subject's viability and therefore rescue or recovery options and urgency. Be sure to clear the peaks of the monitor before entering the space.

Oxygen Deficiency/Oxygen Displacement

OSHA regulations allow entry into atmospheres with 19.5% oxygen and above without respiratory protection. If your monitor indicates less than normal readings of 20.9%, it is absolutely essential to know why they are low. While an atmosphere of 19.5% oxygen may be acceptable in terms of oxygen concentration alone, it may be extremely hazardous due to the presence of other gases and vapors. By thinking of oxygen deficiency as oxygen displacement we will always be thinking that the oxygen may have been displaced by a toxic substance. Approximately 5% of displacing gas is needed to drop the oxygen concentration by 1%. In many cases, 5% of a gas or vapor is very hazardous. The example below uses chlorine in ppm to illustrate this.

1. 1% of volume equals 10,000 ppm
2. 20.9% is the normal oxygen content in air, about one-fifth of the volume
3. If the oxygen content changes to 19.9%, due to 1% being displaced by chlorine gas, there must be a total of 5% of the volume of that gas to also displace the other four-fifths of the total volume
4. That means there is 50,000 ppm of chlorine
5. Chlorine is IDLH at 25 ppm

Components of Air

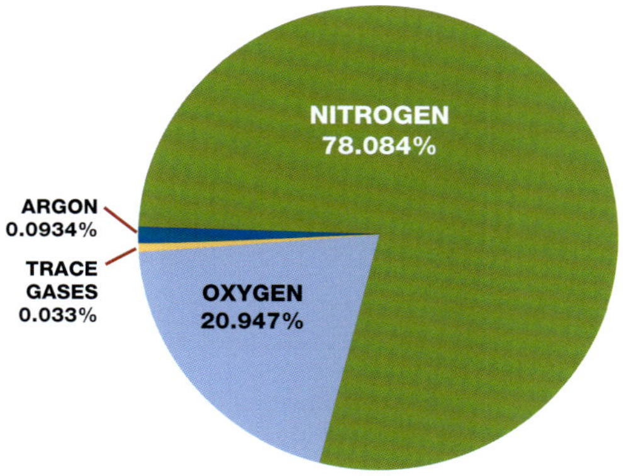

Components of Air with 5% Chlorine

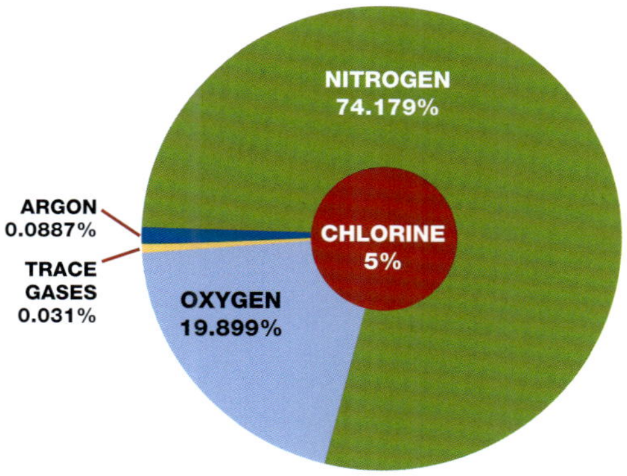

IF YOUR MONITOR READS LESS THAN 20.9% OXYGEN, YOU MUST FIND OUT WHY.

Stratification of Atmospheres

The atmosphere in a confined space, like the atmosphere outside the space, is made up of a combination of gases. Each of these gases has its own weight called vapor density. The vapor density of the gas and its relationship to the vapor density of air will determine the appropriate level to test for that gas.

For example, air has a vapor density of 1.00, and H_2S has a vapor density of 1.1912. An H_2S monitor used only in the top portion of a confined space may not register any significant levels of H_2S. There could actually be a toxic level of H_2S at the bottom of the space because it is heavier than air. If the space was completely filled with H_2S, or if there was sufficient air movement to keep the H_2S dispersed throughout the space, then it is possible to get an H_2S reading at the top of the space.

Just opening a door or hatch of a confined space can cause enough air movement to disperse the toxic contaminant throughout the space. Temperature can also affect the location of toxic contaminants. When cold, what are normally considered lighter than air contaminants may become heavier than air and sink to the bottom of the space.

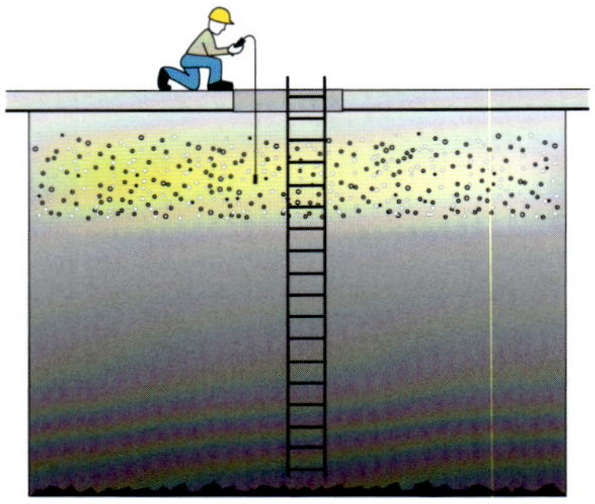

Contaminant with vapor density less than air

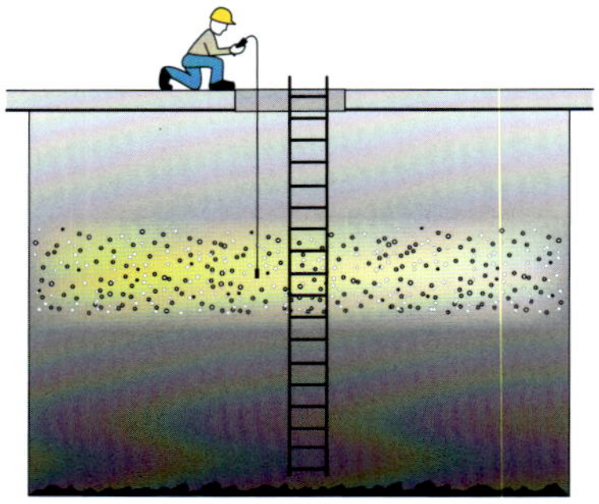

Contaminant with vapor density approximately the same as air

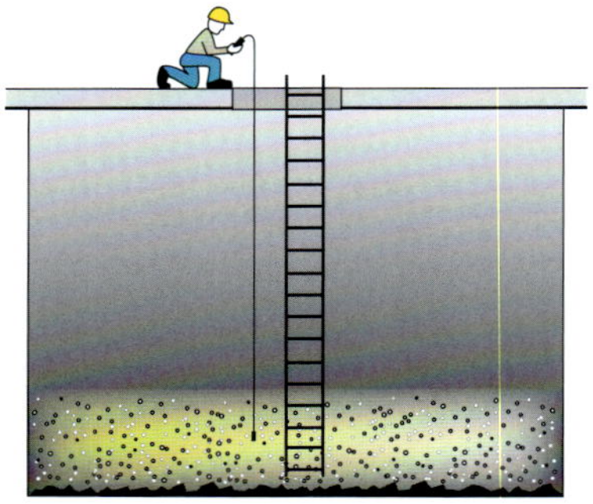

Contaminant with vapor density greater than air

According to OSHA, Appendix B(4), *Testing Stratified Atmospheres,* when monitoring for entries involving a descent into atmospheres that may be stratified, the atmospheric envelope should be tested at a distance of approximately 4 feet (1.22 m) in the direction of travel and to each side. If a sampling probe is used, the entrant's rate of progress should be slowed to accommodate the sampling speed and detector response.

Vapor Densities of Common Gases

Hydrogen	H_2	0.0695
Methane	CH_4	0.5540
Carbon Monoxide	CO	0.9680
Air		1.0000
Hydrogen Sulfide	H_2S	1.1912
Carbon Dioxide	CO_2	1.5270
Sulfur Dioxide	SO_2	2.2638

CHAPTER SIX

HAZARD CONTROL

SCOPE:

This chapter serves as an introduction to hazard control.

TERMINAL LEARNING OBJECTIVE:

At the conclusion of this chapter individuals will be able to describe the most common methods used to control hazards in confined spaces.

ENABLING OBJECTIVE:

- Describe the equipment used to ventilate confined spaces
- Describe ventilation techniques used for various configurations of confined spaces
- Describe common ventilation problems and ways to avoid or overcome those problems
- Describe the ways to lockout/tagout/blockout systems
- Describe the common ways to lockout/tagout/blockout piping systems
- Describe claustrophobia and the methods used to control its effects
- Describe the signs and symptoms of heat illness

Controlling or mitigating confined space hazards is essential to ensure as safe an entry as possible into a confined space for work or rescue. After the assessment has identified the hazards, work quickly to control them.

Ventilation

In the previous chapter we learned that hazardous atmospheres were the leading cause of deaths in confined spaces. Therefore, mastering ventilation techniques is essential for rescue personnel. Ventilation is the most effective means of mitigating atmospheric hazards in confined spaces. Proper ventilation can achieve the following:

- Replace contaminated air with clean, breathable air.
- Decrease the chance of an explosion by keeping the atmosphere below the LEL within the space.
- Reduce/eliminate the toxicity within the space by decreasing the ppm of any toxic substance.
- Increase the chance of survival of any subject(s) trapped in the space by creating a survivable atmosphere.
- The cooling effect of Positive Pressure Ventilation (PPV) is a potential benefit to both subjects and rescuers.

Types of Ventilation

There are two types of ventilation: natural or mechanical (forced).

Natural Ventilation

This relies on the natural motion of the air currents, without assistance, to ventilate the confined space. Depending on the configuration of the confined space, natural ventilation, together with eliminating the introduction of new contaminates, may be enough to control or mitigate atmospheric hazards. At best, this is a slow process and not useful in rescue situations.

Mechanical Ventilation

This type of ventilation uses mechanical means to move air into and out of confined spaces. It is intended to rapidly move a large volume of air to make the space temporarily safe for entry. Mechanical ventilation normally takes the form of either supply ventilation, where you push fresh air into a space under positive pressure, or exhaust ventilation where you are removing the contaminated air. These supply and exhaust methods can be used independently or together depending on the desired results, as each is designed to accomplish different things. A thorough understanding of each of these methods is necessary to safely and efficiently ventilate confined spaces.

Methods of Mechanical Ventilation

Supply Ventilation

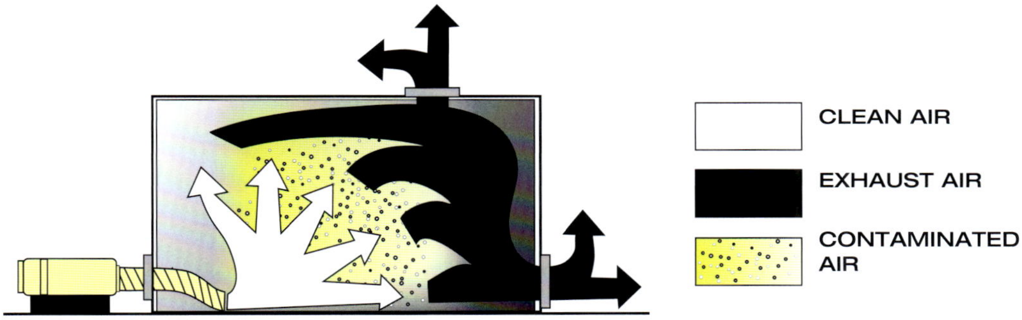

PPV pushes air into the space, causing the contaminated air to exit through any available openings. PPV is used when the contaminated air consists of vapors diffused throughout the space. Tests have shown this to be the most effective type of ventilation, and PPV should be used in conjunction with the other types of ventilation when they are indicated.

Exhaust Ventilation

Exhaust, or negative pressure ventilation, pulls contaminated air out of the space. Fresh air is drawn into the space through any available opening to replace the expelled air. If the exhaust intake can be placed close to the source of the containments or where they are most concentrated, this method can be more efficient than supply ventilation. For example, when the contaminants are heavier than air and the fan intake is placed in the vapor layer at the bottom of the space.

Care must be taken not to contaminate the area outside the space with materials pulled from inside of the space. It is also possible to draw contaminants into the vicinity of the subject and rescuer from other areas of the space.

Exhaust ventilation is considered to be the better way to ventilate flammable or toxic atmospheres. The expelled atmosphere should be ducted away from the working areas around the space, and properly rated fans must be used if the atmosphere is flammable.

General Ventilation

Both supply and exhaust ventilation are classified as general, or dilution ventilation. The contaminated air is diluted by incoming air before being expelled.

General ventilation works well when oxygen deficiency is the main hazard. When used during hazard-producing operations, such as hot work, atmospheric testing or continuous monitoring is required, and the use of respirators may be needed together with general ventilation.

For general ventilation to be effective, the contaminants must not be highly toxic, their concentrations must be low and they must be generated at a fairly uniform rate. The workers should not be close to the contaminant's source.

General ventilation flushes the atmosphere by supplying or exhausting large volumes of air. It moves the contaminants out of the confined space and into the atmosphere, but it doesn't reduce the total amount of contaminants released into the space.

Local Exhaust Ventilation

Another type of ventilation is local exhaust ventilation. This method is used when the contaminants are located or generated at one or more specific points. A well-designed local exhaust system can capture nearly all the contaminants before they are diffused into the atmosphere within the confined space. This is accomplished by placing the exhaust intake at or near the point(s) of origin or concentration of the contaminates.

Local exhaust ventilation is the best way to control flammable and toxic materials generated within the space. Use local exhaust ventilation during hot work, grinding operations or cleaning with solvents.

Sometimes the location or configuration of the space makes local exhaust ventilation impractical. Also, when the operation disperses the contaminant, as in spray painting, general exhaust ventilation should be used.

Combination Systems

Combining supply and exhaust ventilation can increase the efficiency of the ventilation system. This requires multiple fans and ducts.

Ventilation Devices

There are several different devices used to remove contaminated air or supply fresh air to confined spaces.

Fans

Fans are the most common devices used to move air. They are sometimes called blowers. Cost, efficiency and required maintenance may influence the choice of a fan. There are two types of fans, based on the way the air is moved by the fan: centrifugal or axial flow.

Centrifugal Flow Fans—The air enters the centrifugal flow fan parallel to the shaft and is then turned 90° and discharged perpendicular to the shaft. There are several fan blades, each suited for different uses.

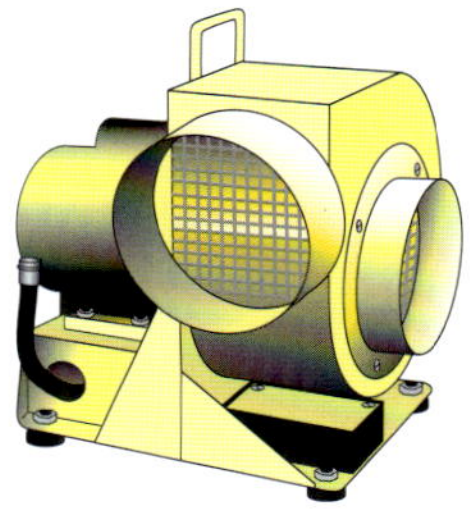

Centrifugal Flow Fan

Paddle wheel (radial blade)—Flat-blade fan. Used for high pressure, medium volume and speed applications. Works well for air with particulate matter.

Forward curved blade—Many narrow, curved-blades set in a shroud ring. Not suitable for air containing particulate matter or other substances that can clog the narrow blades.

Backward curved blade— Flat or curved backward blades that form part of the rotor. Used for large volume, high-speed air flow applications. Will not clog easily.

Axial Flow Fans—Air enters and exits the axial flow fan parallel to the shaft in a straight flow-through design. They are generally smaller and lighter than centrifugal fans that move the same amount of air. There are several different fan blade configurations.

Axial Flow Fan

Simple-axial flow (propeller)—Two- or three-blade propeller mounted on the shaft. Used for moving a large volume of air at low velocities. Does not produce sufficient pressure for use with duct work.

Tube-axial flow—A propeller fan mounted within a cylinder. Moves a large volume at medium pressure, depending on the diameter of the fan and the power of the motor.

Vane-axial flow— Similar to a tube-axial flow fan but with vanes to direct the air in a straight line. Relatively lightweight and small compared to the amount of air it will move.

Ejectors

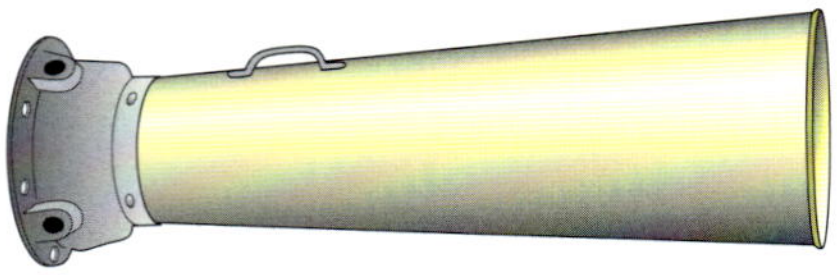

Ejector

Sometimes called jet-air movers, ejectors use the Venturi principle to move air. There are two types.

Air ejectors—Air is blown through a tube creating a low-pressure area within the tube. This causes large quantities of air to be moved through the tube. Air ejectors can be used for supply or exhaust.

Steam ejectors—Similar to air ejectors, but steam is blown through the tube. Steam ejectors can only be used for exhaust.

Ejectors are lightweight, portable and easily connected to duct work. They work well in hot or explosive environments and where contaminates would clog fans. They require large amounts of air or steam and must be grounded or bonded to surrounding structures because of the static electricity that is generated by the moving air. Many air supplies designed to power pneumatic tools inject oil into the supplied air to lubricate the tools. This oil must be filtered out of the air when the ejectors are used to supply ventilation. Ejectors are also extremely noisy.

Ductwork

Ductwork is used to contain and direct the air stream. Different types of duct can be used depending upon the application and what is available.

Flexible, non-collapsible hoses or tubing—May be made of treated fabric or flexible metal. Can be used on either the intake or discharge side of the fan. Short, straight lengths of duct are preferred to maximize air flow, but gentle sweeping curves can be used as needed to conform to the space requirements.

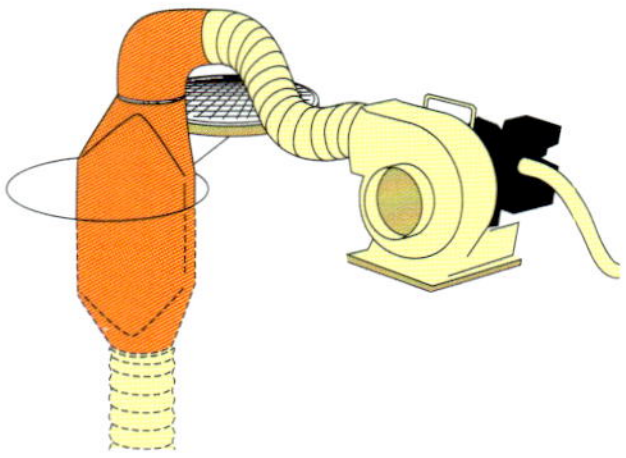

Saddle Vents® and rigid bends—When access/egress must be made through the same opening used for ventilation, a Saddle Vent can be used. Saddle Vents are shaped to provide the largest possible opening for access while at the same time maximizing air flow through the duct. Rigid bends, because of their smooth interior, also help keep the friction losses to a minimum.

Collapsible hoses—Sometimes called a bag duct. May only be used on the discharge side of the fan because they require air pressure to maintain the shape of the duct. The treated fabric or plastic duct is useful in places where personnel must press against the duct to enter, exit or work in the space.

Ventilation Considerations

The factors listed below must be considered when ventilating confined spaces:

Type of atmosphere—Governs the type of fan, whether you use supply, exhaust or a combination system.

Volume of air to move—Determines the size (capacity of the fan) required. Ventilation fans are rated by how many Cubic Feet per Minute (CFM) of air they discharge. This rating may be calculated in free air or through a given length of duct and number of bends (known as effective blower capacity). Using longer lengths of duct and/or more bends adds resistance to the air flow and will result in a lower CFM discharge. Calculate the approximate volume of the space using the following formula: **height x width x depth = volume.**

Typical Fan Tag

Shows atmospheres this fan is rated to operate in.

Shows CFM that this fan will discharge with no bend in the the duct, with one 90° bend in the duct, and with two 90° bends in the duct.

Then divide the volume of the space by the CFM of the fan to determine how long it will take to provide one complete air exchange. When doing this calculation the numbers are usually approximations, so always round up on the volume of the space and round down on the CFM to provide a little more safety.

For example, we are preparing to rescue a worker from a sewage lift station. The approximate size of the space is 8 feet by 8 feet by 6 feet high (8 x 8 = 64 x 6 = 384 cubic feet). Round that up to 400 cubic feet and divide it by the volume of the fan. If we use a fan that discharges 980 CFM and round that down to 950 we will get one air exchange in about half a minute. An alternative method is to use a nomograph such as that found on page 6-8.

Length of time ventilation is needed—This is an area where there is some disagreement. A common industrial ventilation standard when working in flammable or toxic atmospheres is to make five complete air exchanges prior to entry. Others advocate seven or even 10 air exchanges. This is intended for work entry, when the entry time can be scheduled and may not be practical for rescue. Ventilate as long as possible before entry and use your monitoring equipment. It will be necessary to ventilate the space at least long enough to bring the atmosphere to acceptable entry conditions based upon the protective equipment available.

Estimating Approximate Purge Times
100-1,000 Cubic Space Capacity

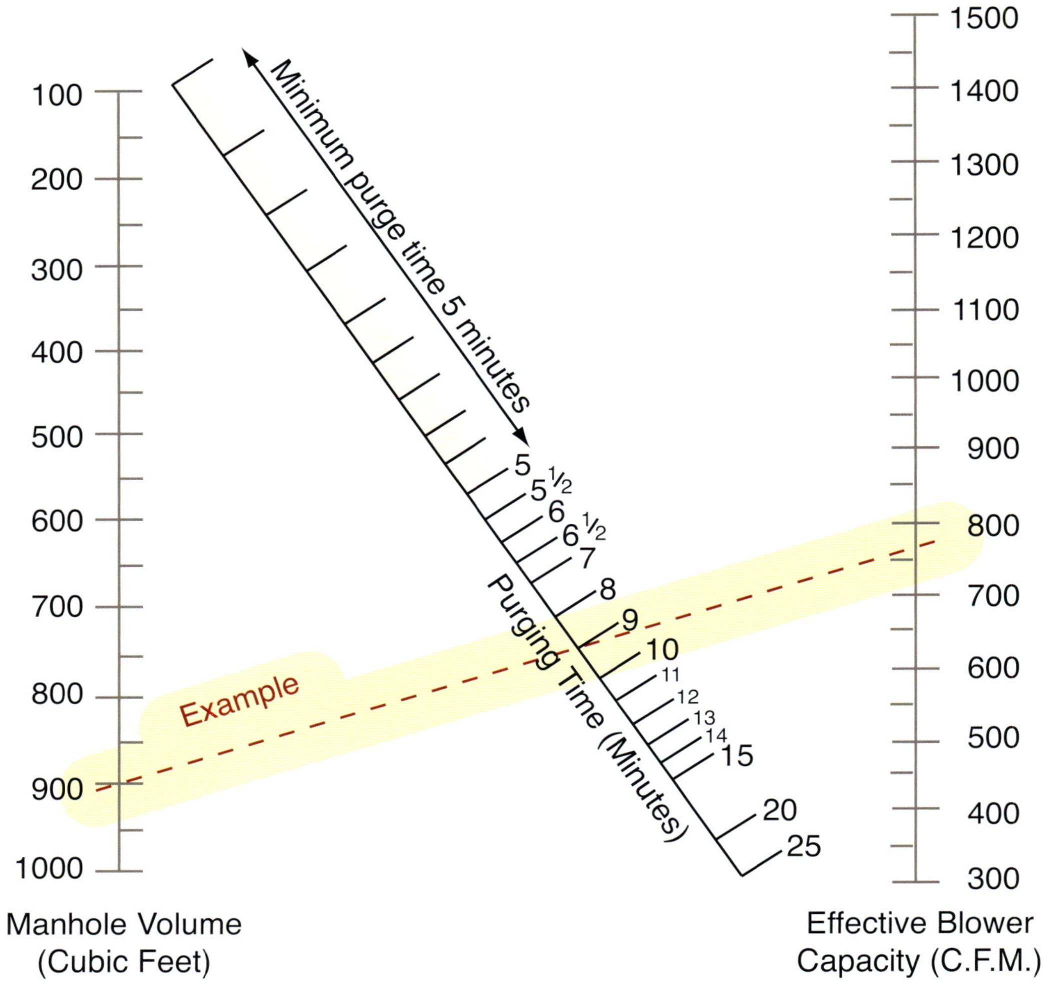

Use of Alignment Chart

1. Place straight edge on manhole volume (left scale).
2. Place other end of straight edge on blower capacity (right scale).
3. Read required purging time, in minutes, on diagonal scale.
4. If two blowers are used, add the two capacities then proceed as above.
5. When toxic gases are encountered, increase purging time 50%.
6. Effective blower capacity is measured with one or two 90° bends in a standard 15-foot blower hose.

REFERENCE: BELL SYSTEMS STANDARD ISS 10, SECTION 620-140-501

Example: A 760-CFM fan with a manhole of 900 cubic feet has a nine-minute purge time.

Estimating Approximate Purge Times 1,000-10,000 Cubic Space Capacity

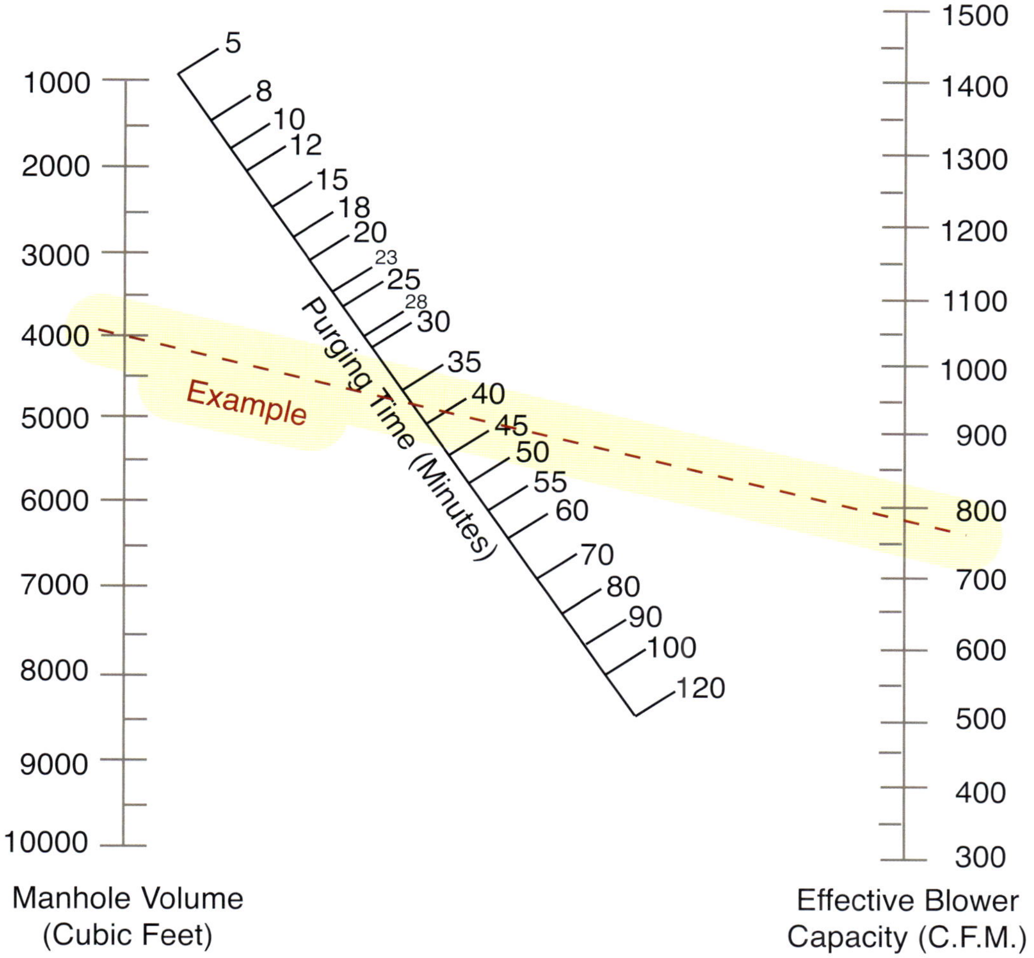

Use of Alignment Chart

1. Place straight edge on manhole volume (left scale).
2. Place other end of straight edge on blower capacity (right scale).
3. Read required purging time, in minutes, on diagonal scale.
4. If two blowers are used, add the two capacities then proceed as above.
5. When toxic gases are encountered, increase purging time 50%.
6. Effective blower capacity is measured with one or two 90° bends in a standard 15-foot blower hose.

REFERENCE: BELL SYSTEMS STANDARD ISS 10, SECTION 620-140-501

Example: A 780-CFM fan with a manhole of 4000 cubic feet has a 36-minute purge time.

Location of fresh air— Governs the amount of duct needed to get clean air into the space. Keep fan intake away from all vehicle exhaust and exhaust ventilation outlets.

Access to space— Determines how ventilation will be routed into and out of the space and the subsequent need for special equipment, such as a Saddle Vent.

Shape of space— Influences the type of directional devices and the air pressure needed to adequately ventilate the space.

Power requirements and availability— Influences fan motor size and power source. Due to the very nature of confined spaces, care must be taken not to negatively impact the atmospheric condition with the device chosen to provide ventilation. For this reason, gasoline-powered fans are not recommended. Evidence has shown carbon monoxide levels can quickly rise within confined spaces ventilated by gasoline-powered fans. The exhaust from the engine can be pulled into the air intake and subsequently blown into the space. Even if ducts are used to adequately separate the air intake from the exhaust, personnel working around the fan remain exposed to exhaust fumes.

Pneumatic and hydraulic-powered fans are available and work well if these power sources are available. Electric fans are rapidly becoming the standard as they can be powered by local electrical power or generators located away from the rescue site. When used as exhaust ventilation in flammable atmospheres, electric fans must be approved for the actual or potential atmosphere within that space. Rated fans are available that have fully enclosed motors and switches and anti-static containers and ducting. There is not a standard for the fan assembly, only for the electrical components, so it is the user's responsibility to know the classification of the location or environment and to determine whether or not the fan is safe to use under the actual circumstances.

Note: Ventilation must be stopped during initial atmospheric monitoring to get a true sample of the atmosphere within the space. When ventilation is on, the sample may be artificially diluted by the incoming air.

Ducted Ventilation Problems

There are some problems that must be avoided when setting up a ventilation system.

Recirculation

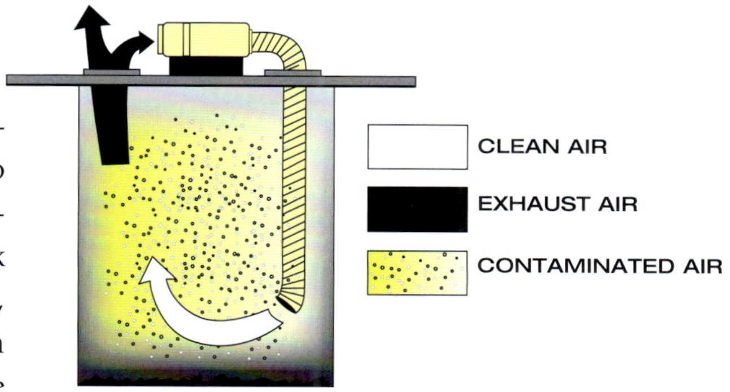

This happens when contaminated air is exhausted too close to the air inlet, recirculating the contaminated air back into the space. To prevent this, position the inlet away from the exhaust. Keep the inlet five feet away and upwind of the opening. A duct may need to be run to the clean air source.

Short Circuiting

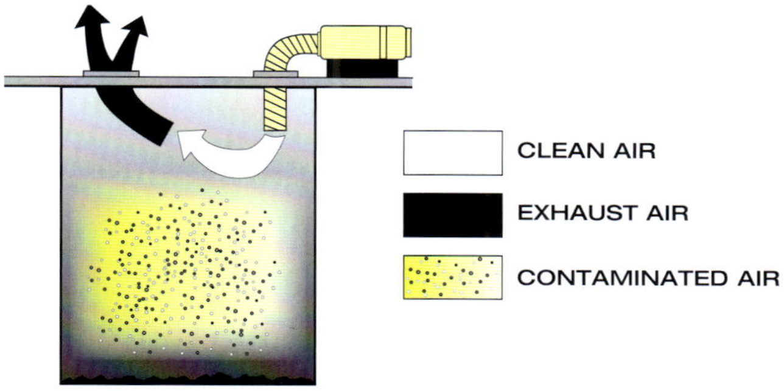

This happens when air moves directly from the air inlet to the exhaust outlet without circulating through the other areas of the space, leaving them unventilated. Extend ducts far enough into the space to force air into all areas of the space. Use walls as diffusers to direct air into corners and all levels.

Chimney Effect

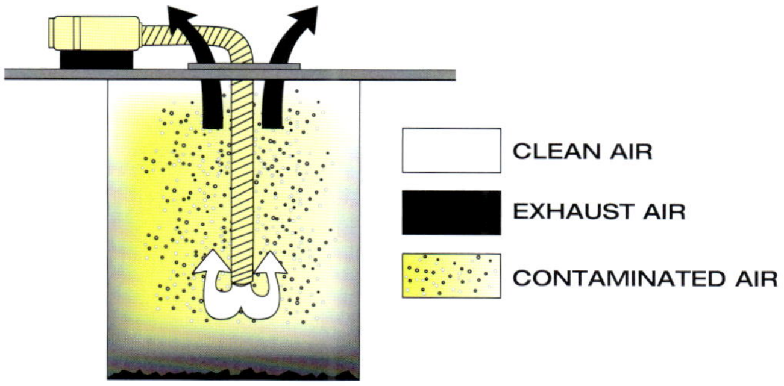

This occurs when the duct is hung straight down into a vertical opening and the fan is not powerful enough to force air into all areas of the space. As the air loses force leaving the duct, it follows the path created by the duct and flows back out of the space. Position the end of the duct near the corner of the space to use the walls to diffuse the air, causing a turbulent, ricochet effect that will purge all areas of the space. Use a fan with enough power to move the air throughout the space.

Specific Ventilation Procedures

OXYGEN DEFICIENCY

In situations where there is an oxygen deficiency, caused either by consumption or a contaminent, supply ventilation is the best way to furnish oxygen.

Openings at Either End of a Long Space

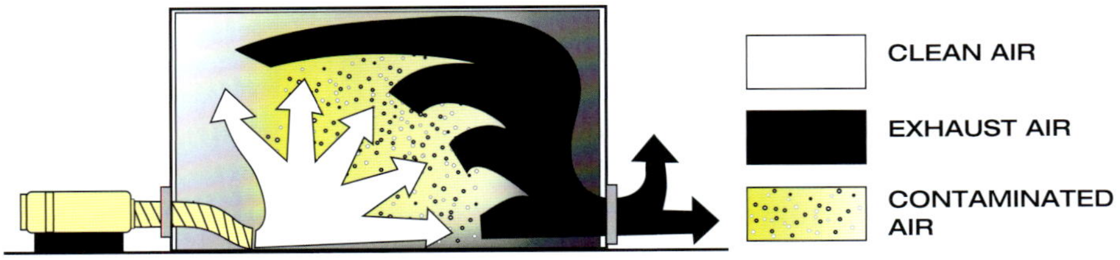

Set up the blower to supply fresh air at one end of the space and exhaust the stale air out the far end. This gives the best distribution of fresh air, and recirculation of the exhaust air is unlikely because the openings are so far apart.

Spaces with Only One Opening

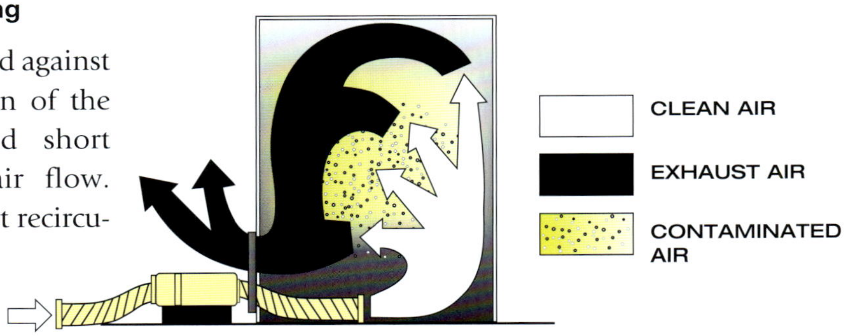

You must guard against both recirculation of the exhaust air and short circuiting the air flow. To protect against recirculation, place a duct outside the space to protect the air supply from the exhaust stream. To protect against short circuiting, use a blower with enough velocity to ventilate the entire space.

Deep Confined Space

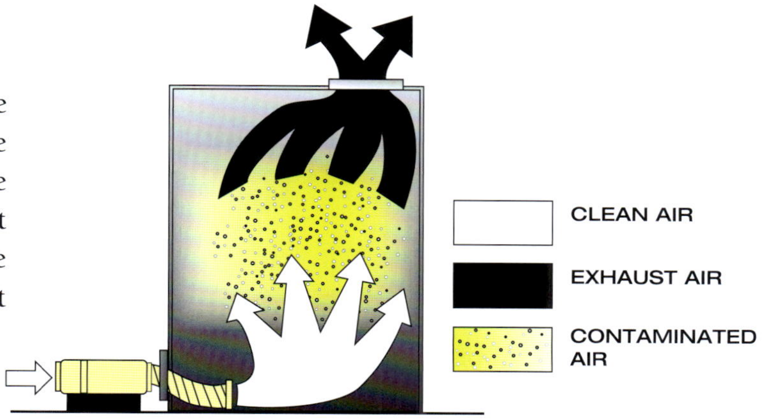

Introduce the fresh air at the bottom with the exhaust outlet at the top. If there is no opening at the bottom, use ductwork inside to direct the air to the bottom of the space. The blower must operate at a high velocity to provide good mixing and the best exchange of good air for stale air.

FLAMMABLE OR TOXIC CONTAMINANTS

Exhaust ventilation is best for ventilating flammable or toxic contaminants because it removes the contaminant without dispersing it throughout the space.

Long Spaces with One Opening at the End

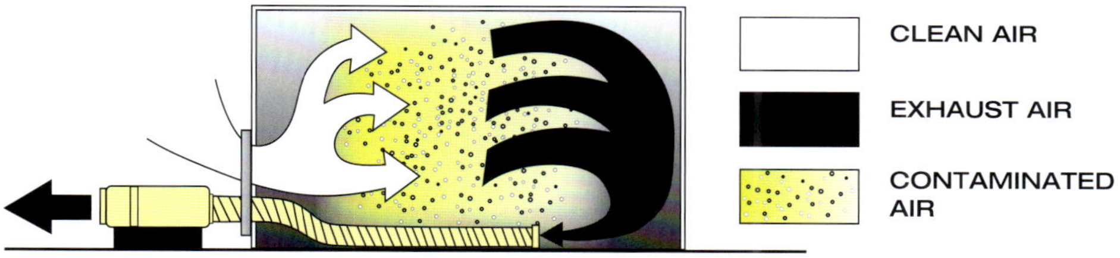

Use ductwork to place the exhaust fan inlet at the far end to draw fresh air through the space.

Heavier-Than-Air Contaminant

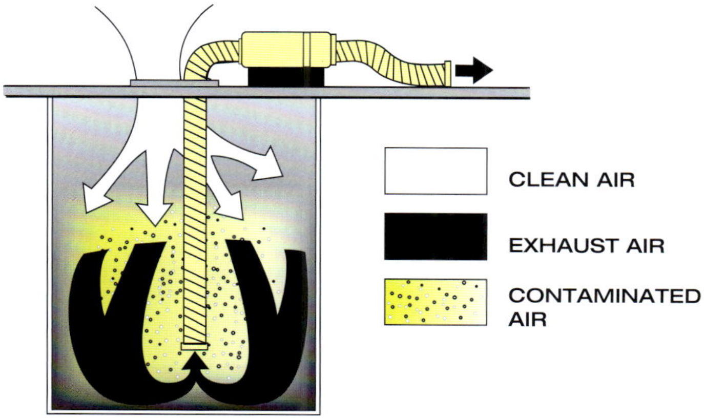

Place the exhaust fan inlet near the bottom of the space if the atmosphere remains stratified. In spaces with only a top opening, extend the duct down to the level of the contaminant.

Lighter-Than-Air Contaminant

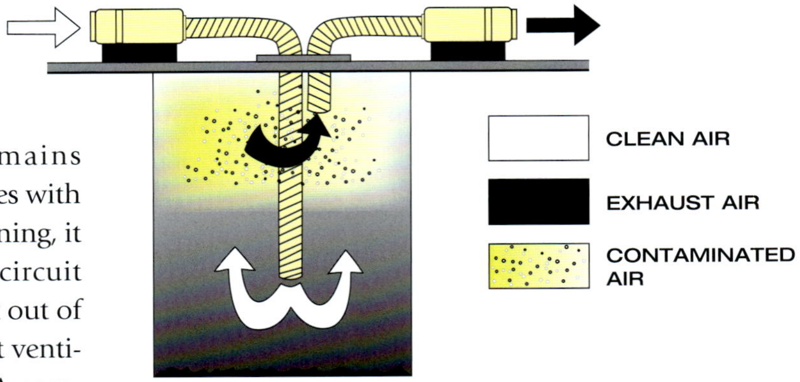

Place the exhaust inlet near the top of the space if the atmosphere remains stratified. In spaces with only one top opening, it is easy to short circuit the good air back out of the space and not ventilate other areas. A combination system with supply air ducted down to the bottom of the space will work better.

Single Point of Origin

Local exhaust ventilation works best. Keep the exhaust inlet close to the source of contamination to reduce the chance that contaminants will escape. The airflow velocity must be high enough to capture and carry the contaminants.

Local Supply Ventilation

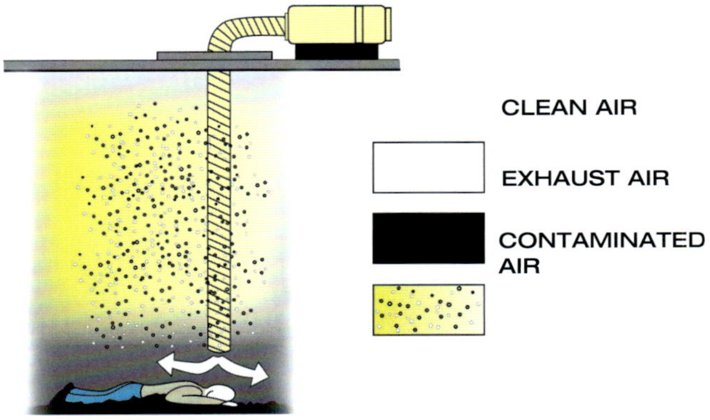

Local supply ventilation can be used in situations where the subject is accessible from the opening of the space and it is possible to place the end of the supply duct directly on the subject's face. This technique, though unproven, provides an area of fresh air around the subject while other preparations are being made for an entry rescue. The benefits of supply ventilation are still achieved while getting fresh air to subject sooner than would otherwise be possible.

Non-Ducted Ventilation Techniques

Ducted fans provide the most versatile ventilation, but sometimes you must make do with what is available. Fans without ducts can be used to ventilate confined spaces. Electric box fans are typically found in the fire service. They can be used for both supply and exhaust ventilation.

Spaces with Two Openings

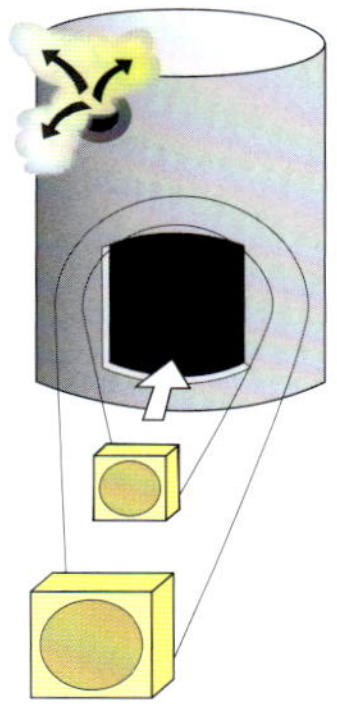

Place the fan far enough back from one opening that a cone of air seals the entire opening, pushing fresh air into the space. Use the largest opening for the supply side if possible. The exhaust opening should be on the opposite side of the space or downwind of the supply fan to prevent recirculation. Multiple fans can be placed in series, one blowing on the other, to increase the velocity of air being moved into the space.

Spaces with One Opening

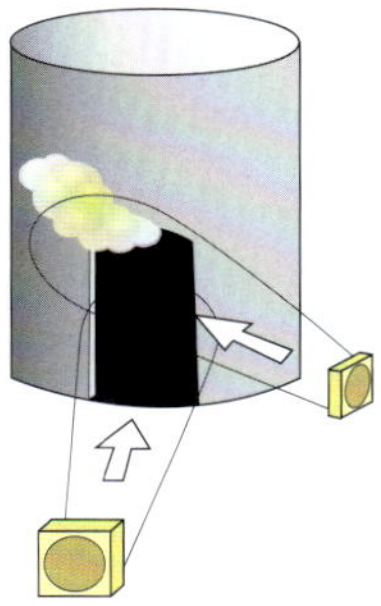

When there is only one opening, move the fan close enough to only partially cover the opening. This allows the contaminants to exhaust through to the rest of the opening. Place a second fan at 90° to the opening and far enough away that it won't disrupt the first fan but will blow the contaminant away. This prevents recirculation and protects personnel working around the opening.

Important Safety Considerations When Ventilating

REMEMBER THESE SAFETY CONSIDERATIONS BEFORE YOU BEGIN VENTILATING:

- Are fans compatible with the environment you are ventilating? (Explosion proof for flammable atmosphere.)
- Are all vehicles and generators positioned downwind and away from any openings into the space and ventilation inlets to assure exhaust fumes are not pulled into the space by your ventilation efforts?
- Begin ventilation far enough in advance to allow sufficient time to bring the atmosphere into acceptable entry conditions.
- Only ventilate with clean air, never use pure oxygen. Pure oxygen can create a serious fire hazard.

- Continually monitor the atmosphere to assure that you are not moving into an explosive condition by adding oxygen to a too-rich environment.
- Use flagging tape or other "telltales" to monitor air movement in the space and warn of ventilation failure.
- Ensure that no debris can be sucked into the supply ventilation fan. Place the fan or inlet duct on a tarp, salvage cover, plywood or other clean hard surface.
- Make arrangements for a second power supply to ensure continuous ventilation should the original power supply fail.
- Make every attempt to monitor the space prior to ventilating to establish a baseline.
- Direct exhaust away from the entrance to the space and other work areas. This is especially important when ventilating contaminated air.

Atmospheric Condition		Ventilation
Oxygen Deficiency	Dispersed in Atmosphere	General Supply (PPV)
Flammable Atmospheres	Dispersed in Atmosphere	General Exhaust
Flammable Atmospheres	Concentrated (single point of origin)	Local Exhaust
Toxic Atmospheres	Dispersed in Atmosphere	General Exhaust
Toxic Atmospheres	Concentrated (single point of origin)	Local Exhaust
Hot Work	Concentrated (single point of origin)	Local Exhaust
Grinding Operations	Concentrated (single point of origin)	Local Exhaust

Controlling Hazardous Energy

Control of hazardous energy is a key component of rescuer safety in confined space operations. OSHA regulates this subject in 29 CFR 1910.147. That performance-based regulation went into effect on January 2, 1990. It is commonly referred to as lockout/tagout but is really much more than that. OSHA defines hazardous energy control as:

> "a program of energy control procedures and employee training to ensure that before any employee performs any servicing or maintenance on a machine or equipment where the unexpected energizing, start up or release of stored energy could occur and cause injury, the machine shall be isolated and rendered inoperative."

It is important to note that this OSHA standard is designed to eliminate electro-mechanical hazards only. Hazards such as those created by flowable products, such as steam, natural gas, hydraulic fluid, fuels, sewage, etc., into a permit-required confined space are considered eliminated only by usage of the techniques described in the definition of the term "isolation" in paragraph (b) of 29 CFR 1910.146 and CCR 5157. Isolation means the process in which the confined space is effectively removed from service and completely protected against the release of energy and material into the space. Isolation may be accomplished by blanking or blinding, misaligning or removing sections of lines, pipes or ducts, a double block-and-bleed system or disconnecting all mechanical linkages. Before using a double block-and-bleed system (one in which a pipe is closed by two valves and vented to the atmosphere between them) check the regulations. Some jurisdictions and industries have banned its use. Engulfment is the second biggest killer in confined spaces, and many accidents could be prevented through compliance with a comprehensive lockout/tagout program, which insures that all potentially hazardous equipment is rendered safe and placed in a "zero mechanical state (ZMS)."

The Six Steps to Lockout/Tagout

1. Preparation for shutdown— Know the types and magnitude of energy involved, the hazards of the energy and the methods of controlling the energy.
2. Shutdown—This may be as simple as throwing a switch or it may require a more orderly, methodical shutdown sequence.
3. Isolation— Locate ALL energy isolation points that control the particular equipment involved.
4. Application of lockout/tagout devices —lockout/tagout devices shall be installed by an authorized employee to hold the energy isolation points in a safe position.
5. Control stored energy— Assure all potentially hazardous stored energy is controlled and/or alleviated.
6. Verification of isolation—An authorized employee shall verify, prior to entering the space or performing work on equipment, that it has been locked and/or tagged.

It should be pointed out that in a rescue situation the rescue team will probably not have the time, expertise or equipment to conduct major lockout/tagout operations. They were called because something went wrong during a confined space entry and the accidental release of stored or hazardous energy or material is one more thing to consider to help ensure rescuer safety.

Energy Sources

Electrical—Electrical energy is the most widely used source of energy and is most often associated with lockout/tagout procedures. Electrical energy is any electric current, AC or DC, used to power equipment, machinery or control systems.

Pressure—Pressure is used as power and power assist in various types of heavy equipment and to assist in valve actuation. Pressurized systems apply force to move or work equipment. The pressurizing medium can be instrument air, or in some cases, a gas in pneumatic systems. In hydraulic systems, oil-based or synthetic fluid may be used. Hydraulic jacks and positioning equipment, crane boom systems, pneumatic valves and steam systems may contain pressure energy.

Note: Process fluids, although under pressure for transport, are not defined as pressure energy sources.

Momentum/Gravity—Many pieces of equipment utilize momentum or gravity, the force of motion contained in moving equipment caused by gravity or inertia to carry out their function. This is especially critical to identify once the power source has been shut down. Electrical energy may be used to raise a press and gravity is used to lower the equipment. Pumping unit counterweights, compressor drive and flywheel systems and grinder wheels may have momentum or gravity energy.

Residual/Stored Energy—This is power that may be left over after the power has been shut down, shut-off or disconnected. This energy may not be obvious and may be used to hold equipment in place. Springs, capacitors, valve actuators, compressor cylinder chambers, gas-filled shocks and struts may all contain residual or stored energy.

Note: If a device has residual or stored energy, it is considered energized until such a time the residual or stored energy is controlled.

Lockout/Tagout/Blockout

The purpose of a lockout/tagout/blockout system is to control the following energy sources:

- Electrical sources within the space, such as conduit or electrical wiring.
- Mechanical items, such as augers, mixers, conveyors, doors, etc.
- Any item that may release energy or begin movement or action while personnel are in the space. It is necessary to identify those items in which stored energy potential is possible and assure that all stored energy is either released or neutralized.

All mechanical devices and equipment capable of causing injury shall be placed in a ZMS.

Electrical

Electrical lockout/tagout

All electrical equipment (excluding needed lighting) shall be locked in the open (off) position with a key or combination padlock. The key shall stay with the person who placed the lock. Off-site rescue teams should have their own locks to put in place with those who did the lockout. The key can be kept by the person who placed it or a designated person, such as the safety officer.

The appropriate tag must be placed on the electrical equipment that has been locked out. The tags must give the reason for placing the tag, the identity of the person placing the tag, how that person can be contacted and the date and time the tag was placed. Off-site rescue teams can have tags printed with department name and telephone number, giving rescue as the reason for the tagout and be marked with the date and time when it was placed.

If the equipment is not capable of being locked, a tag shall be securely attached to the equipment with a non-reusable self-locking device (nylon cable tie) capable of enduring at least 50 pounds of pull. Assigning a stand-by person to assure that the equipment is not accidentally turned on is the best way to protect rescuers.

Note: Equipment manufactured after January 2, 1990, must be designed to accept lockout devices.

Mechanical

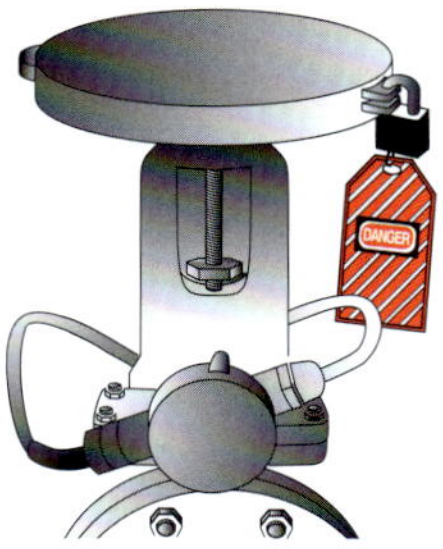
Valve with wheel cover device

Valves in hydraulic or pneumatic actuating lines must be locked in the closed position and any pressure bled off. The valves can be locked with a valve wheel cover device or with a chain. Mechanical equipment that can inadvertently move by sliding, falling or rolling must be blocked. Cribbing and wedges can be used in place of special brackets or stands made for this purpose. Coiled springs, spring-loaded devices or suspended loads must also be released, so that their stored energy will not result in inadvertent movement.

Pipes or Ducts

Blinding or blanking procedures are designed to eliminate the possibility of any substance entering the space by a pipe or duct system. Blinds or blanks are disks of metal placed in a pipe or duct that completely occludes it to ensure that nothing can enter the space. There are several different type of blinds used in pipe flanges. Be careful not to mistake a spacer, screen or orifice plate as a blind. Disconnecting and misaligning the pipe will also prevent product from entering the space, but a release outside the space may still endanger the rescuers. The double block-and-bleed system, where two in-line valves are closed and locked out and the pipe between them is bled to the outside, is also allowed by the regulation. Be aware that many industries do not allow it to be used due to the high incidence of valve failure. The three types of blinds most commonly used today are:

Slip Blind—A rated and sized metal plate inserted between gasketed pipe flanges to prevent the flow of gas or liquid in either direction.

Spectacle Blind—A combination blind and spacer formed from one piece of material. Frequently, a permanent part of the line, the open or closed ends are switched depending on whether or not the flow is required.

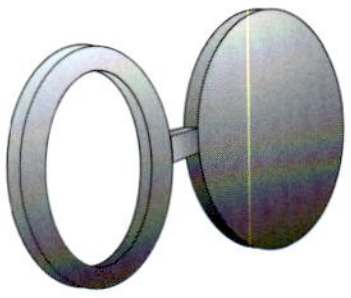

Blind Flange—A full-rated pipe fitting used to close the flanged end of an open pipe or valve.

DO NOT CONFUSE BLINDS WITH SPACERS, SCREENS OR ORIFICE PLATES.

Spacer

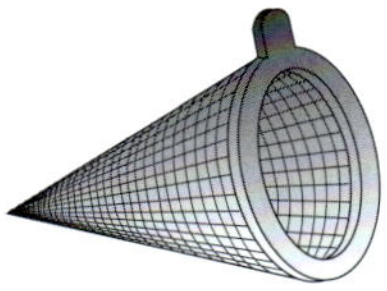
Screen

Orifice Plate

When an off-site rescue team is responding to an industrial rescue it is important to remember the following:

- Industrial processes can be very complex. Attempting to isolate a confined space without complete knowledge of the entire system can be disastrous. Stopping the flow of a substance into a vessel can have serious consequences back down the line that not only can endanger the rescue operation but needlessly cause expensive repairs. In compliance with lockout/tagout regulations, each industry has developed energy control plans that deal with isolating each confined space on the facility. Finding a qualified representative of the site and reviewing the energy control plan as you go over the lockout/tagout that has been performed and placing your own locks and tags as you go is the only safe way to ensure isolation of industrial confined spaces.
- Stored energy might still be present and will need to be controlled.
- Backup or redundant systems in an industrial setting may override one system and bring another on line. This means that if one system is shut down a backup comes on. It is necessary to control all systems that may effect the confined space.
- OSHA requires that a person who locks out or tags out machines or equipment in order to perform servicing or maintenance on that machine or equipment be an authorized employee.

The key to a successful lockout/tagout or blind system is retaining someone intimately familiar with the electrical and mechanical systems in the area, plant or space where you are making entry. Allow these personnel to brief and guide you on their systems!

Ignition Prevention

When entering a potentially explosive atmosphere, do not transport an ignition source into the atmosphere. All equipment taken into an explosive atmosphere must be considered a potential ignition source. Are the tools used to mitigate the hazard non-sparking? Are the radios that you use for communications safe? How about that camera that you are about to use for that close-up picture for the incident commander, is it safe? All of these need to be considered prior to making an entry.

But how do we know that electrical equipment is safe? There are standards set, and organizations, such as UL, FM, ANSI and NFPA, have developed test protocols for certifying devices to meet minimum standards of acceptance. Certification means that a device is certified to prevent a potential ignition source from igniting a flammable atmosphere.

An electrical device certified under one of these test methods must carry a permanently affixed plate showing the logo of the laboratory granting certification and the class, division and group it was tested against. A device that is certified and is used, maintained and serviced according to the manufacturer's instructions will not contribute to ignition. Remember, the device is certified for use only in the specific atmospheres for which it was tested.

There are methods used to make these devices safe. The two most common methods used to make devices safe are explained below.

Explosion Proof

Explosion-Proof Light Label

This method encases the ignition source in a rigidly built container. These instruments allow the flammable atmosphere to enter. If and when an arc is generated, the ensuing explosion is contained within the specially designed and built enclosure. Within it, any flames or hot gases are cooled prior to exiting into the ambient flammable atmosphere so that the explosion does not spread into the environment.

Intrinsically Safe

This method reduces the potential for arcing among components by encasing them in solid insulating materials. Also, reducing the instrument's operational current and voltage below the energy level necessary for ignition of the flammable atmosphere provides equal protection. An intrinsically safe device, as defined by the National Electrical Code, is incapable of releasing sufficient electrical or thermal energy under normal or abnormal conditions to cause ignition of a specific hazardous atmospheric mixture in its most easily ignited concentration. Abnormal conditions shall include accidental damage to any wiring, failure of electrical components, application of over-voltage, adjustment and maintenance operations and other similar conditions.

Claustrophobia

Claustophobia is the confined space hazard that you bring with you. Controlling this hazard is mainly up to the individual, but other entry team members and the attendant need to be able to recognize the signs if the individual misses the symptoms or is unable to control them.

Confrontation Techniques

- This begins by entering and confronting the confined space.
- Expect established avoidance patterns.
- Remember that avoidance only reinforces the phobia. Failure to confront the fear will increase the stress levels at future events. You must trust your team, your equipment and yourself!
- Peer pressure and personal feelings will play a part in your thought process; accept this.
- By knowing these reactions are normal, it is possible to expect them and deal with them.
- Should you fail to accept these normal feelings you may progress to a state where your sense of unreality is brought on by imagined horrors (e.g., the ground collapsing, air running out, walls closing in).

Control Techniques

Gaining an understanding of how to recognize these signs and symptoms is the beginning of control. Consider the following process for dealing with phobias.

- **Expect, allow and accept the oncoming fear**.
- **Don't fight pre-entry feelings**.
- **Be short-range, goal oriented**—Deal with the pre-entry and entry in stages, concentrating on each one as it presents itself.
- **Suiting up**—Assure your personal protective gear is all in order and working, checking each piece of gear. Check your buddy and confirm with him.

- **Approaching**—Get to the entrance and concentrate on your movement into the space, including any entry methods that will be used, rope gear, tripods, etc.
- **Entry**—Concentrate on making a safe and efficient entry, coordinate your moves and assist your teammate. Focus on a safe and smooth entry process. Once inside, focus on short-distance goals, each a little longer. Talk with your teammate and coordinate your efforts in reaching one point after the other, as if following a compass course.
- **Concentrate on the rescue, not your fear**—Accomplish your assigned primary task.
- **Expect the phobic wave**—Recognize it for what it is and use your knowledge to your advantage.
- **Communicate, talk, laugh, joke, use the buddy system**.

Medical Monitoring

Medical monitoring is often overlooked during emergency situations. Monitoring the fitness of the rescuers (before, during and after the rescue) is a important factor in preventing medical emergencies involving rescue personnel. If medical monitoring is done in a consistent manner, it will establish a baseline for the rescuer and will help determine if the rescuer is physically fit enough to be allowed to attempt another rescue.

Body temperature is an area that can affect the rescuer's ability to be effective in a rescue. In a confined space with limited air circulation, the heat buildup from internal and external sources can cause a dramatic difference from the outside ambient air temperature. Normal body temperature is 98.6°F. In a hot, humid environment, such as a confined space, the rescuer involved in extreme physical exertion will run the risk of raising the core body temperature. As the core temperature raises, several things will happen.

SIGNS AND SYMPTOMS OF HEAT STRESS/FATIGUE

- Body core temperature range 99.5°F-101.3°F
- Confusion/poor judgment
- Loss of coordination
- Chills

SIGNS AND SYMPTOMS OF HEAT EXHAUSTION

- Body core temperature range 101.3°F - 105°F
- Confusion/poor judgment
- Loss of coordination

- Chills
- Decreased level of consciousness
- Pale, cool, sweaty skin
- Headache and extreme weakness
- Nausea and vomiting
- Dizziness
- Profuse sweating
- Weak, rapid pulse
- Rapid, shallow breathing

In severe cases, the body core temperature rises above 105°F. In this range the body would go into heat stroke. This is a true life-threatening emergency and requires immediate medical attention.

SIGNS AND SYMPTOMS OF HEAT STROKE

- Body core temperature above 105°F
- Decreased level or loss of consciousness
- Hot, dry skin
- Rapid pulse
- Rapid, shallow breathing
- Hypotension

The body's other vital signs (pulse, blood pressure and respiration) also need to be monitored and recorded. The pulse needs to be monitored for signs of heat exhaustion or stroke and pulse irregularities. Pulse rate is a good indicator for determining if the heart and body are at an acceptable level for the type of strenuous work that is done in a confined space rescue. Blood pressure is also a good indicator in assessing if the heart and body have received enough rest and are in condition to work under strenuous conditions. Monitoring of the rate and quality of respirations can assess several things. Increased respiration can indicate overheating, overexertion or possibly be signs of phobic reaction to the confined space. Monitoring of lung sounds for any congestion or fluid in the lungs after the operation may indicate an inhalation exposure.

Note: The three stages of heat-related illness (heat stress/fatigue, heat exhaustion and heat stroke) can overlap, and the symptoms of more than one stage can be present at the same time.

SUGGESTED STANDARDS FOR MEDICAL MONITORING[1]

Remove from work any person manifesting any of the following signs:

Temperature: > 100.4° F.

Heart Rate: > 85% of the maximum heart rate (Maximum heart rate = 220 minus your age)

Heart Recovery Rate: > 10 beats per minute drop after 1 - 3 minutes of rest

To determine heart recovery rate, take pulse 1 minute after exiting space. Take pulse again 3 minutes later. If the heart rate hasn't dropped at least 10 beats per minute, remove person from work.

Respiration: Congestion, wheezes or other respiratory difficulties

NO PERSON SHOULD BE RETURNED TO WORK UNLESS CLEARED BY A MEDICALLY QUALIFIED AUTHORITY.

[1]Standard set forth by the California Specialized Training Institute and the California State Fire Marshal for Medical Monitoring referring to hazardous materials entries.

CHAPTER SEVEN

PERSONAL PROTECTIVE EQUIPMENT

SCOPE:

This chapter serves as an introduction to personal protective equipment.

TERMINAL LEARNING OBJECTIVE:

At the conclusion of this chapter individuals will be able to describe the personal protective equipment necessary to enter and safely work in confined spaces.

ENABLING OBJECTIVE:

- Describe the physical protection required to safely enter and rescue a subject from a confined space
- Describe the four levels of chemical protective equipment
- Describe Self-Contained Breathing Apparatus (SCBA) used in confined spaces
- Describe supplied air respirators and escape cylinders used in confined space rescue
- Describe air purifying respirators and their advantages and disadvantages when used in confined space rescue
- Describe respiratory protection factors and how they are used to determine respiratory protection levels
- Describe the elements of a respiratory protection program as required by state and federal regulations

Physical Protection

One of the most important aspects of a confined space rescue operation is the safety and survivability of confined space rescuers. Rescuers typically enter and work in dangerous and hostile environments that may be oxygen deficient or enriched, contain a flammable gas, contain toxic hazards, be dark, noisy, have temperature extremes and be just large enough for one rescuer to wiggle into. PPE needs to provide an acceptable level of protection from the hazards of the confined space, be lightweight and allow the rescuer to move freely. The type of space being entered and the environmental and physical hazards associated with the confined space should all be considered when choosing the following PPE for rescuers.

Helmet—The ideal helmet for confined space rescue is lightweight, has a low profile with a small brim, has top and side impact resistance, has an adjustable three-point suspension system, can accommodate a helmet mounted lighting system and is approved by ANSI. The typical structural firefighting helmet is too big and heavy and has a very large profile that makes it difficult to enter and work in very small confined spaces, and it contributes to rescuer head and neck fatigue. This helmet does a great job of protecting firefighters during fire-related incidents but is not well suited for confined space rescue work.

Flash Protection—Rescuers should not enter a known flammable atmosphere, but once rescuers enter the space atmospheres may suddenly change and a minimal level of protection from a flammable atmosphere is desirable. Rescuers are required to continuously monitor confined space atmospheres and if a flammable atmosphere is detected, rescuers will need to exit the space as soon as possible. Personal protective clothing that is made of fire-resistant material, has long sleeves, is reinforced at the elbows and knees and will provide suitable rescuer protection from the physical hazards of confined spaces. Structural firefighting turnout clothing is not recommended for use in confined space rescue. Turnout clothing is bulky, restricts rescuer movement and contributes to rescuer heat stress and fatigue. A hood that covers the head, neck and ears and is made of a fire-resistant material will provide a minimal level of protection from fire and heat related hazards in a confined space.

Boots—Generally, leather safety boots with a steel toe and a high ankle that provides adequate rescuer ankle support are suitable foot protection for confined space rescue work. The boot soles should have an aggressive tread that allows good footing in slippery, wet and uneven terrain. Because leather will absorb most liquids, including liquid chemical hazards, chemical-resistant boots may be necessary if the confined space environment contains a liquid that could cause a chemical exposure to the feet.

Gloves—Gloves need to be heavy enough to provide an adequate level of hand protection but lightweight so as to allow maximum rescuer dexterity when tying knots and manipulating equipment. Lightweight leather work gloves that have a snug fit and provide protection from abrasive and sharp hazards are a good choice for confined space rescue work. Leather gloves may also provide some protection from flammable hazards. Another type of hand protection is a two-part glove. When worn together, the inner fire-resistant glove and the outer leather glove provide a higher degree of fire and abrasion resistance. The inner glove is constructed of a thin fire-resistant material that allows the rescuer to tie knots and manipulate equipment with some dexterity while providing a degree of flash protection. While the two-part glove provides additional protection, having to take gloves on and off in a confined space can delay a rescue operation and increases the potential for losing a glove.

Elbow and Knee Protection—Elbow and knee protection may be required when working in confined spaces that are so small that rescuers need to crawl on their hands and knees, or in even smaller spaces, while wiggling through the space on their abdomen and the tip of the toes. Just as we would never enter and work in a confined space with only a single layer of fire-resistant material covering our feet, rescuers should always protect their knees and elbows from rough, sharp and abrasive hazards. Elbow and knee protection that is padded and provides protection from sharp and angulated objects and has a degree of impact resistance should always be considered.

Ear and Eye Protection—Eye protection is required any time the eyes may be exposed to any type of hazard. Goggles provide excellent all-around eye protection from dust, dirt, debris and liquid splash hazards. Protective glasses and face shields provide limited protection from flying objects and do not provide protection from flying dust or objects that may enter the eyes from the bottom or sides.

Atmospheric Monitor—Rescuers should carry an atmospheric monitor with them when working in a confined space and continuously monitor and assess the atmosphere in which they are working. A monitoring instrument that is capable of testing the atmosphere for oxygen, flammability and toxicity is required. Atmospheric monitoring equipment as well as other electrical devices used in the confined space need to be certified as safe for the atmosphere in which they are used, so that they do not become a source of ignition. Atmospheric monitors and hazardous atmosphere ratings are discussed in greater detail in Chapter 5, Atmospheric Monitoring.

Lighting—Rescuers will also need to have some type of lighting equipment available to illuminate dark environments. As with atmospheric monitoring instruments, lighting equipment also needs to be certified as safe for the hazardous atmosphere in which it will be used. Rescuers should carry three types of lighting equipment with them: a helmet-mounted lamp, a hand light or flashlight and a chemical light stick. Helmet-mounted lamps free the rescuers hands and illuminate the area the rescuer is looking at. A flashlight is carried as a backup light in case the helmet lamp fails and chemical light sticks are carried as a "cold" light source that will not be a source of ignition in a flammable atmosphere.

Communication—The success or failure of a confined space rescue operation is dependent on whether clear and effective communication takes place between rescuers. Effective communication should be established by using the minimum amount of equipment that gets the job done. Electrically operated equipment used in confined spaces also needs to be certified as safe for the hazardous atmosphere it will be used in. There is a more detailed discussion of communications equipment and methods in Chapter 11.

Tag Line—A tag line may be taken into the confined space by the first entrants and used as an aid to quickly find their way out of the space in a hurry. A tag line may also be used to mark the location of the subject or, if the subject is not found, the furthest point reached. The tag line can be used by subsequent rescue teams to quickly locate the subject or the furthest point reached or to locate entrants who may need assistance in exiting the space. A SAR airline may substitute for the tag line.

Chemical Protective Equipment

Confined space rescuers typically enter confined spaces to rescue someone who has been overcome by the hazards associated with the product previously stored within the space. Even though the confined space was emptied, residual product may remain and create an exposure hazard to rescuers. The previously stored contents of the space need to identified, the hazards of the space eliminated/controlled, the atmosphere tested and monitored and the appropriate rescuer PPE selected and worn. The type of chemical hazard in the space and how the chemical enters the body (route of entry) will determine what level of protection is required. There are four ways a chemical hazard can enter the human body:

- Ingestion
- Injection
- Inhalation
- Absorption

The United States Environmental Protection Agency identifies four levels of personal protective equipment ensembles:

LEVEL A

- NIOSH-approved SCBA or supplied air respirator with escape SCBA
- Fully encapsulating vapor-proof chemical protective suit
- Chemically resistant inner gloves
- Chemically resistant outer gloves
- Chemically resistant boots with steel toe and shank
- Two-way communication

LEVEL B

- NIOSH-approved SCBA or supplied air respirator with escape SCBA
- Chemical-resistant clothing (overalls and long-sleeved jacket, coveralls, hooded two-piece chemical splash suit, disposable chemical-resistant coveralls)

- Chemically resistant inner gloves
- Chemically resistant outer gloves
- Chemically resistant boots with steel toe and shank
- Two-way communication

LEVEL C

- Full-face or half-mask, approved air purifying respirator
- Chemically resistant clothing
- Chemically resistant inner gloves
- Chemically resistant outer gloves
- Chemically resistant boots with steel toe and shank

LEVEL D

- Structural fire-fighting turnout clothing
- No respiratory protection
- Should **not** be worn in areas where respiratory or skin-exposure hazards exist

Rescuers will need to determine the route of entry of the chemical hazard and take steps to protect that route of entry. If the chemical hazard can be absorbed through the skin, then a Level A ensemble is required. If the chemical hazard presents an inhalation route of entry, then respiratory protection is required. Level A ensembles are big, bulky, restrict rescuer movement and vision and require specialized training. For these reasons, rescuers should not enter confined spaces wearing Level A ensembles. OSHA requires that hazards be eliminated/controlled from confined spaces prior to entry and chemical PPE worn only if all other methods of eliminating hazards do not adequately protect rescuers.

The ideal personal protective clothing ensemble for confined space rescuers is a lightweight jumpsuit or jacket and pants made of fire-resistant material with long sleeves and reinforced at the elbows and knees. If a liquid splash hazard exists (that does not present an absorption hazard), then gloves and boots impervious to liquids should be worn.

Respiratory Protection

The NIOSH Alert, *Preventing Occupational Fatalities in Confined Spaces,* concludes that "...deaths in confined spaces often occur because the atmosphere is oxygen deficient or toxic." For this reason rescuers should be very careful if they enter confined spaces without wearing some sort of positive-pressure self-contained or supplied air respiratory protective equipment.

Selection of Respiratory Protection

Respiratory protection is of primary importance since inhalation is one of the major routes of exposure to chemical toxicants. Respiratory protective devices consist of a face piece connected to either an air source or an air-purifying device.

Respirators with an air source are called atmosphere supplying respirators and consist of two types of open-circuit units: **self-contained breathing apparatus (SCBA)** and **supplied air respirators (SAR)**. SCBAs usually consist of a face piece connected by a hose and a regulator to a bottle of compressed breathing air carried by the wearer. These bottles supply from 30 to 60 minutes of air, depending on the size and pressure of the bottle. This is the minimum form of respiratory protection recommended for entry in to an IDLH atmosphere. SCBAs have a limited supply of air, are bulky, heavy and increase the chance of heat stress.

Another form of SCBA is a closed-circuit rebreather. This unit reuses the air that has been exhaled by removing the carbon dioxide with a chemical reaction. Rebreathers are designed for up to four hours of use. If an SCBA supplies clean air from a cylinder and then exhausts the exhaled air to the atmosphere, it is considered an open circuit. When it recycles exhaled gases, it is considered a closed circuit.

SARs have air supplied from a source located a distance away and connected to the user by an air-line hose. NIOSH states that the hose shall be of breathing-type quality and no longer than 300 feet in length. Individual models vary and some may not be rated to 300 feet. Check the manufacturer's specifications to be sure the unit meets your needs.

SARs cannot be used alone in an IDLH atmosphere. If something happens to the hose or air supply, the rescuer's lives would be endangered. For safety, an escape bottle must be worn in conjunction with an SAR. These bottles normally hold five to 15 minutes of air and can be turned on if a failure in the main air supply occurs.

SCBAs and SARs are further differentiated by the type of air flow supplied to the face piece. Negative-pressure respirators, also known as demand respirators, draw air into the mask when a negative pressure is created by the wearer's inhalation. The disadvantage of negative pressure is that if any leaks develop in the system, the user draws contaminated air into the face piece during inhalation. Positive-pressure respirators maintain a positive pressure in the face piece at all times. These are typically pressure-demand respirators with a pressure regulator and an exhalation valve on the face piece. Only positive-pressure respirators should be used in emergency response.

Self Contained Breathing Apparatus (SCBA)

Typical backpack-style SCBAs are available that will accommodate 30-, 45- and 60-minute air supply bottles. Greater air capacity equals greater weight; greater weight equals greater stress and fatigue to the rescuer. SCBA worn on the rescuer's back creates a larger rescuer profile that may limit or prevent entry through small portals and limit rescuer movement inside the space. SCBA is designed to be worn on the back. **It is very important that rescuers not remove backpack-style SCBA to enter and/or work inside a confined space.** Removing SCBA from the back and passing it through confined space openings is dangerous. Rescuers can easily lose control of their SCBA, causing the face mask to be pulled off and exposing the rescuer to a hazardous atmosphere. SCBA contains a finite amount of air. Air management and controlling rescuer operational time inside confined spaces becomes critical when conducting rescue operations using SCBA. The rescue group supervisor, the attendant and the entry team members need to agree on the rescuer operational time inside the confined space. Operational time must be based on the type of space, the environmental conditions inside the space, the workload involved in packaging and removing the subject from the space and a safety factor (the amount of air we want the rescuers to exit the space with). It is considered poor air management to use the SCBA low-alarm device as an indication of when rescuers should exit the space.

SCBA may be equipped with a supplied air connection allowing the rescuer to breathe air supplied by an air hose connected to an air source outside of the space. The SCBA bottle then becomes an escape air supply and functions much like a SAR.

Even with the limitations of SCBA, it may be appropriate in some situations for rescuers to enter and work in confined spaces wearing SCBA. The following guidelines should be considered when using SCBA for confined space rescue operations:

- Enter through an entry portal without removing SCBA from the back
- Work in the space without removing SCBA from the back
- The subject should be located near the entry portal
- The attendant should be able to keep the rescuers within sight
- Drag subject to the entry portal for packaging so an emergency exit from the space can be made quickly.

Another solution to the finite air supply of SCBA would be for the entrants to take extra air into the confined space with them. Several SCBA manufacturers make a carrying strap and regulator that attaches to an air bottle, similar to a rapid intervention team air pack, that can be carried into the space and connected to a rescuer SCBA equipped with a dual air-line connection.

Supplied Air Respirators (SAR)

Supplied air respirator systems are designed specifically for use in confined spaces. SAR systems consist of:

- Harness
- Face mask
- Regulator
- Emergency escape bottle
- Breathing air hose
- Air source and regulator

Supplied Air Respirator
with Escape Bottle

Breathing air is supplied to the rescuer through an air hose connected to an air source outside of the space. NIOSH requirements limit the air hose length to 300 feet from a regulated air source. The air source can be an air cart or manifold that holds two or more air bottles connected to a regulator. One bottle is opened and used until the low-pressure alarm device sounds and then a second bottle is opened. A check valve in the manifold prevents the full bottle from backfilling into the empty bottle.

An emergency escape air bottle is attached to the harness and provides an emergency air supply to the rescuer should the main air supply be disrupted. OSHA regulations require emergency escape air bottles (auxiliary self-contained air supply) to be worn with SAR when working in IDLH atmospheres. Typical emergency escape air bottles range in size from five to 15 minutes. The rescuer should exit the space immediately once the emergency escape air bottle is opened. The emergency escape air bottle should never be used to extend the rescuers reach beyond the length of the supplied air hose.

Another type of air source for a SAR is a mobile or fixed breathing air unit with an air compressor. Breathing air is supplied to the rescuer through air hose connected to a stored air system that is recharged by a compressor. When the stored air supply pressure becomes low, the air compressor makes more breathing air. These types of breathing air systems allow rescuers to potentially operate indefinitely inside confined spaces.

Air Purifying Respirators (APR)

Air Purifying Respirator

Air purifying respirators are not connected to an external air source and only filter the air in the atmosphere. Air is inhaled through air filters or canisters that remove chemical contaminants from the atmosphere. Filters and cartridges are available for a variety of dusts, vapors and mists. The atmospheric hazard must be identified and the proper filter or cartridge selected. Cartridges also have a service life and must be replaced periodically. Cartridges can become saturated quickly in atmospheres containing high concentrations of chemical contaminants and should not be used in IDLH atmospheres. It is important to remember that APRs only filter chemical contaminants from the atmosphere.

There are two types of APRs. With a **non-powered APR**, the user causes the air to move through the filtering system by causing a negative pressure in the mask. **Powered air-purifying respirators (PAPR)** utilize a blower to assist the user in breathing and producing a positive pressure in the mask. The advantages of APRs are their small size and light weight, which have little effect on the rescuer's mobility. They are also easy to maintain.

The following guidelines must be considered when using APRs:

- Atmospheres must contain at least 19.5% oxygen. The atmospheric contaminant must be identified
- The atmosphere (including the contaminant) must be monitored continuously
- Do not use APR in IDLH atmospheres
- The APR must be equipped with an End-of-Service-Life Indicator (ESLI) certified by NIOSH for the contaminant
- If there is no ESLI appropriate for conditions, a change schedule for canisters and cartridges, based on objective information or data, that ensures that canisters and cartridges are changed before the end of their service life must be implemented

Respirator Protection Factor

The level of protection that can be provided by a respirator is indicated by the respirator's protection factor. This number is the ratio of the concentration of a contaminant outside the face piece relative to the concentration inside the face piece. ANSI recommends that SCBAs with full face piece in positive pressure have protection factors of 10,000. This means that a team member wearing such a respirator should be protected in an atmosphere containing chemicals at concentrations that are up to 10,000 times higher than the appropriate limits. Protection factors are assigned to respirators by NIOSH.

Respiratory Protection Program

Employers who have employees that are required to wear respiratory protection in the workplace are required to comply with the appropriate respiratory protection regulations and "develop and implement a written respiratory protection program with required worksite-specific procedures and elements for required respirator use. The program must be administered by a suitably trained program administrator."

The written respiratory protection program should address the following:

- Procedures for selecting respirators for use in the workplace
- Medical evaluations of employees required to use respirators
- Procedures for proper use of respirators in routine and reasonably foreseeable emergency situations
- Procedures and schedules for cleaning, disinfecting, storing, inspecting, repairing, discarding and otherwise maintaining respirators
- Procedures to ensure adequate air quality, quantity and flow of breathing air for atmosphere-supplying respirators
- Training of employees in the respiratory hazards to which they are potentially exposed during routine and emergency situations
- Training of employees in the proper use of respirators, including putting on and removing them, any limitations on their use and their maintenance
- Fit-testing procedures for tight-fitting respirators
- Procedures for regularly evaluating the effectiveness of the program

CHAPTER EIGHT

PHASES OF CONFINED SPACE RESCUE

SCOPE:

This chapter serves as an introduction to the organizational structure of a confined space rescue.

TERMINAL LEARNING OBJECTIVE:

At the conclusion of this chapter, the student will describe the order and phases of a confined space rescue.

ENABLING OBJECTIVE:

- Describe Phase I— Preparation, including rescue team evaluation, equipment evaluation, personnel evaluation, preplan hazard analysis and incident management
- Describe Phase II— Assessment, including approach assessment and resource assessment
- Describe Phase III— Pre-entry operations, including survival profile of the subject, site safety, hazard control and pre-entry briefing
- Describe Phase IV— Entry and rescue operations, including the duties of the rescue group supervisor, attendant, entrants and backup team
- Describe Phase V— Termination, including accountability of personnel, retrieval of personnel and equipment and documentation

In this chapter we consolidate many of the skills learned in other chapters into the organization of the rescue operation. Please note that some of what we call phases occur before the actual rescue operation. The specialized equipment required in confined space rescues as well as the training to use it are things that must be acquired long before the emergency call is received. NFPA 1670 has additional requirements about levels of training by persons performing rescues in confined spaces and the current edition should be consulted by agencies that are required to follow that standard.

It would be impossible to describe the organization of every rescue incident because they are all different. Some might require extensive rigging; others may not require any. Some can be accomplished with out breathing apparatus; others may require long lengths of air line and large quantities of breathing air. Our goal is to provide you with the information necessary to build an organizational structure to manage the incident.

Phase I: Preparation

Rescue Team Evaluation

One of the questions that must be asked and evaluated long in advance of any incident is whether the rescue team is competently trained and equipped to perform a confined space rescue. This does not mean having a rope and an SCBA and thinking you are ready to perform a confined space rescue. Your rescue team evaluation should include the following questions:

- Does the rescue team understand and conform to the OSHA standard as closely as possible?
- Does the rescue team have enough trained personnel to sustain a confined space operation? Four or five trained personnel are not enough to safely complete a confined space rescue.
- Does the rescue team have the proper equipment to perform this operation? Confined space rescues require specialized equipment.

Equipment Evaluation

The following is a minimum equipment list a rescue team should have before considering itself even marginally ready to perform these types of operations.

- SAR—One for each member of the entry and backup teams and one for the subject
- Cascade system capable of on-site delivery of high-volume breathing air. A primary system for the entry team and the subject plus a secondary system for the backup team.

- Patient-packaging equipment compatible with confined spaces
- Patient removal equipment, such as a tripod, davit or cable winch
- Rope, harnesses and hardware
- Atmospheric monitors capable of detecting:
  - ✓ 0_2 levels
  - ✓ Flammability — % LEL of methane or other calibrated combustible gas
  - ✓ Toxicity of gases such as CO or H_2S ppm
- Ventilation equipment
- Explosion proof lighting
- Intrinsically safe communications equipment
- PPE for rescue personnel compatible with confined space operations
- Lockout, tagout, blankout kit
- Personnel accountability and tactical work sheet

Preplanning (Hazard Analysis)

During the initial preparation phase, rescue teams need to consider preplanning for specific hazards that may effect their operations and identify potential rescue sites. This can be done by identifying the following:

- Are there any locations, industries, etc., that routinely perform confined space entry for cleaning or general maintenance?
- Are there any special projects being undertaken that are in confined spaces?
- Does your response area include facilities with high hazard potentials such as water or sewage treatment facilities?

These basic questions will provide you with a general overview of your confined space problem. Contact representatives that work in these areas and request an on-site inspection. If appropriate, ask for copies of prior entry permits or other site-specific information, such as energy control or emergency plans.

- Identify any special hazards, chemicals or processes that need to be known before a response
- Identify special, site-specific problems, such as location, access, air supply problems or extreme elevation differences

Sample Confined Space Survey Form

CONFINED SPACE NUMBER	PERMIT REQUIRED ☐ YES ☐ NO	DATE OF SURVEY

☐ Alternate Entry Procedures Possible
☐ Atmospheric Hazards Only ☐ Permit-Entry Required for Testing ☐ Eliminated by Ventilation

☐ Reclassification as Non-Permit Space Possible
☐ No Actual or Potential Atmospheric Hazards ☐ All Other Hazards Eliminated
☐ Permit-Entry Required for Hazard Elimination

LOCATION OF SPACE

DESCRIPTION OF SPACE

CONFIGURATION OF SPACE

POSSIBLE ATMOSPHERIC HAZARDS
☐ Oxygen < 19.5% ☐ Oxygen > 23.5% ☐ Flammable Atmosphere ☐ Toxic Atmosphere

POSSIBLE GASES PRESENT

PHYSICAL HAZARDS
☐ Mechanical ☐ Electrical ☐ Pneumatic ☐ Hydraulic ☐ Pipes or Ducts
☐ Engulfment ☐ Hazardous Substance ☐ MSDS Required TEMPERATURE: ☐ High ☐ Low

UNUSUAL HAZARDS

REASONS FOR ENTERING SPACE

WHO USUALLY ENTERS SPACE

NUMBER OF ENTRY POINTS	FREQUENCY OF ENTRY

EXTERNAL CONNECTIONS TO SPACE

SURVEY COMPLETED BY

Print Name Signature

A full-size version of this form can be found in the appendix.

Preplanning is essential to any rescue team seriously attempting to provide confined space rescue services. It not only allows incident commanders and rescuers alike to plan the most effective rescue, but it will also increase the safety of the overall operation.

Personnel Evaluation

Just as important as having the right equipment is having the right personnel. Rescue personnel must have the same knowledge and training as work entry personnel in addition to training in rescue operations and equipment. These requirements are listed in the regulation. Training in actual confined spaces can bring to light one of the serious problems that effect rescuers, claustrophobia. Rescuers who cannot control this phobia will not be much help to the patient. Training in related disciplines, such as hazardous materials and ICS, is important. Rope rescue skills used in high-angle rescue will translate directly into the confined space field. Showing competency in all these skills is mandatory before attempting a confined space rescue.

Incident Management

Managing a confined space rescue requires delegation of the many duties necessary for a successful operation. The Incident Command System (ICS) is a framework for managing any type of incident. Is is used by many municipal agencies and industry alike. Prior training in ICS is important so all personnel have a clear understanding where they fit into the incident. This framework must be in place before you can assign duties.

Phase II: Assessment

Assessment is the first operational phase of the confined space rescue incident. Assessment begins from the moment the alarm is sounded and in some ways continues until the completion of the rescue operation. Survival and successful operations at the scene of a confined space emergency are dependent on good information-gathering processes. By identifying a coordinated method in which to gather, interpret and disseminate information, the operation is more likely to end successfully.

Establish the Incident Command System

A management system must be in place as soon as possible to assign assessment duties and allow other personnel to begin preparation for the next phase. This process should be automatic and may vary by organization. The initial incident commander may be the first-in rescue team leader or fire officer. As time goes by or as the incident escalates they may be replaced by a higher ranking officer, company safety manager or whomever is so designated. The important thing is that someone trained and experienced with confined space rescue needs to be in charge and to begin making decisions and assigning the appropriate personnel to the positions that need to be filled in the command structure. Because confined space incidents are of an unusual and, hopefully, rare in nature it is recommended that an incident command chart, such as that shown below, be followed to help trigger the memory of the specialized positions that must be filled.

Incident Command System

The ICS spans all phases of a rescue operation. One of the fears of untrained personnel when they see an ICS organizational chart is that it takes too many people to manage an incident, leaving no one to work it.

Confined Space Rescue Incident Command System

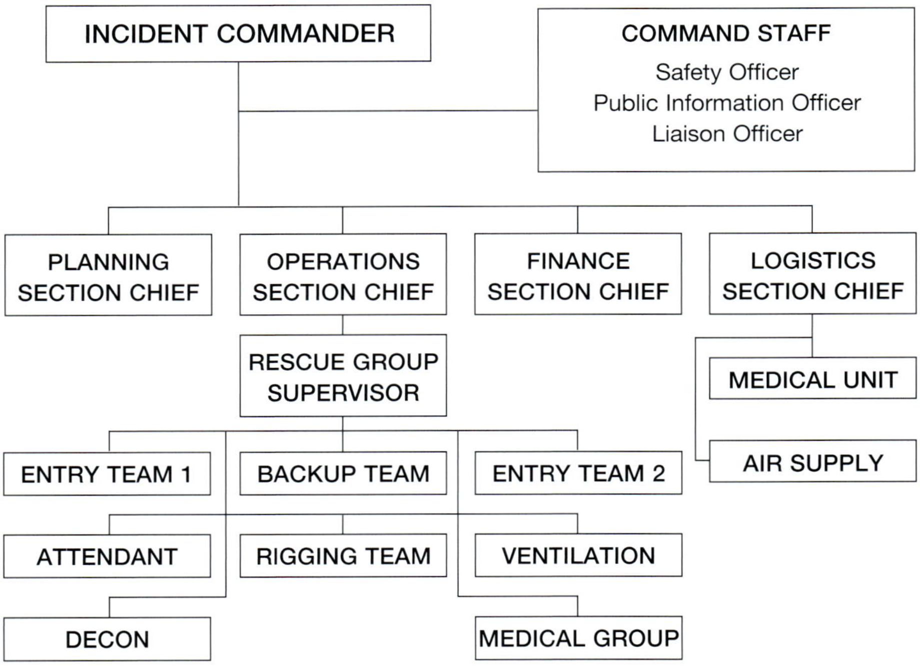

A small incident can be managed by one person wearing many hats. If the incident stays small, then that one person may be the only overhead person. In a larger incident, one person may assume many duties but then delegate some or most of the duties as personnel arrive. The ICS is able to expand as the incident grows to keep a efficient span of control.

Incidents that have more than one agency or organization with responsibilities are managed by a unified command. Representatives from all responsible agencies or organizations establish one set of incident objectives and strategies, while the representative with the most qualifications and experience acts as the incident commander. All agencies retain authority, responsibility or accountability. When off-site rescue teams respond to an industrial rescue where the employer remains responsible for the injured employee, a unified command would be used.

Incident Commander—Responsible for overall management of the incident. Directly controls the command staff and the section chiefs.

Command Staff—Work directly under the incident commander.

Safety Officer—Responsible for developing and recommending measures for assuring personnel safety and to assess and/or anticipate hazardous and unsafe situations. Should not have any other responsibilities to distract him from his primary duty. In large operations, the safety officer may have additional safety personnel assigned to specific areas of the operation who report to him.

Public Information Officer—Responsible for developing and releasing information about the incident to the news media, incident personnel and to other appropriate groups or agencies.

Liaison Officer—The contact for the personnel assigned to the incident by assisting or cooperating agencies not involved with tactical assignments or in a unified command. In cases where an incident occurs at an industrial site and the rescue is performed by off-site rescue teams that provide incident management, the liaison officer is also responsible for working with the site owner or their representative.

Section Chiefs—There are four sections in the ICS.

Planning Section—Responsible for collection, evaluation, dissemination and use of information about the development of the incident and the status of resources. Supervises preparation of the incident action plan. Planning also conducts the briefing and debriefing sessions.

Finance Section—Responsible for tracking expenses of the operation, including equipment, supplies and man-hours. While there is the tendency to overlook this function during the conduct of the mission, it can be very important later. Confined space rescues are very hard on equipment, some of which will require a thorough cleaning if not outright replacement. Proper accounting documentation will be required in cases where billing for costs of labor and/or damaged equipment is allowed.

Logistics Section—Responsible for supplying all resources needed to complete the incident.

Operations Section—Responsible for the management of all operations involved in the incident.

Logistics Section Chief—Responsible for the accumulation, location and distribution of all needed equipment. The following positions are part of this section:

Air Supply—Responsible for all air supply, bottles, supplied air breathing apparatus and any other related items as they apply to the use of SAR, SCBA and the related breathing air supply.

Medical Unit—Responsible for the medical care of rescuers, which includes baseline readings for medical monitoring of the rescue team.

Operations Section Chief—Responsible for the coordination of the operational portion of the incident. Within this section are the following positions:

Rescue Group Supervisor—Responsible for the entire rescue group.

Attendant—Has many of the same duties as for work entry; communicating with the entry team, air monitoring, tracking entrants and equipment.

Ventilation—Responsible for ventilation inside and outside of the confined space.

Entry Team—Performs all duties inside the confined space, such as reconnaissance, patient extrication and packaging.

Backup Team—Ready to make entry if the entry team needs rescue.

Rigging Team—Responsible for the coordination and evacuation of patient and entry teams and operation of retrieval systems. There may be more than one rigging system involved with the rescue operation. One required to gain access to the space, for example, a tripod over an opening at the top of a tank, and another to get the subject from the top of the tank to the ground. If a rigging system is required within the confined space, that will need to be constructed by the entry team.

Decon—Responsible for all decontamination operations.

Medical Group—Responsible for medical care of patient.

Many confined space rescue operations are small because of the size of the environment that the rescue must take place in. One person may assume the combined duties of the IC, PIO, liaison, planning section chief, finance section chief, logistics section chief, operations section chief and rescue group supervisor. As the operation expands and more personnel arrive, he may choose to delegate some of these duties.

Assessment is divided into two areas when dealing with confined space operations.

Approach Assessment

This is intended to act as an initial size-up tool and to identify hazards.

While responding to the incident the rescue team should review any pre-plans associated with the specific confined space or with spaces with similar hazards or characteristics.

Upon arrival at the scene the following questions should be asked and the answers evaluated. This information is crucial to the development of an incident action plan.

While not specifically required, we suggest the use of some sort of confined space rescue permit to assist with jogging the memory for what to ask as well as to serve as the first step in the documentation of the rescue operation.

- What is the main problem?
  - ✓ How many personnel are trapped or injured?
  - ✓ How many personnel are unaccounted for and where were they last seen?
- What type of space is this?
- What is the space used for?
- Is the space currently in use?
- Is this a product storage area?
  - ✓ Are there product storage hazards?
  - ✓ Is there a viscous or heated material?
  - ✓ What residue is possible?
  - ✓ Is there an engulfment potential?
- What other hazards are there in the space?
  - ✓ Electrical, mechanical or stored energy?
- What are the entry and exit points?
  - ✓ Are there multiple entry points?
  - ✓ Are the entry points above ground or below grade?
  - ✓ Are there other access problems?

Once these initial questions are answered, the operation planning can begin.

Resource Assessment

This will identify whether you have adequately trained manpower and resources to accomplish this task either on-site or responding.

After evaluating the information received in the approach assessment and developing an incident action plan, look at the resources on-site and en route to determine if additional resources are needed. Requesting any additional resources immediately gives you the best chance at a successful operation.

Documentation

Prior to moving on to the next phase, it is mandatory to document all actions and assessments made since starting the operation. This documentation must be continued throughout the entire incident. Early and complete documentation serves as both a checklist for all necessary and required pre-entry safety measures, but also becomes a legal record of actions taken. In many discussions with OSHA officials, the one consistent comment made is, "If it's not documented, it didn't happen." This documentation can be in the form of a confined space entry permit (see page 8-21), which employers are required to complete prior to entry into permit-required confined spaces. A confined space rescue permit (see page 8-22) can be developed that works as a checklist for your rescue procedures. A tactical log (see page 8-23) or worksheet that records all activity involved with the rescue, including logging all entry team movement into and out of the space, provides necessary documentation. Be sure to include not just times and actions, but the reasons for using, or not using, specific procedures or resources. This is then filed for later review to evaluate your rescue program.

Responsibilities during the Assessment Phase

Assessment begins when the first reports reach the incident commander before and after arriving on the scene and continues as other personnel arrive. Many of the duties related to assessment will be quickly completed and are not part of the regulated operational job descriptions during the pre-entry and entry phases.

Incident Commander

- Set up a logical and safe command post.
  - ✓ Upwind and away from the incident.
- Assign key positions as needed.
  - ✓ Safety officer should be assigned early in the operation.
  - ✓ Other permanent positions can be assigned or temporary duties can be tasked to available personnel and the permanent assignments can be made when more trained personnel arrive on-scene.

- Locate and retain a responsible person from the incident site.
  - ✓ This should be someone familiar with the operation and isolation procedures of the space.
- Establish control of the rescue site.
  - ✓ Only allow personnel with PPE into areas that have not been cleared of hazards.
  - ✓ Place personnel, vehicles, cones or tape to establish a perimeter to keep unwanted people and vehicles away.
- Atmospheric monitoring outside and inside the space.
  - ✓ The attendant may be a good choice for this assignment. His duties in latter phases will include controlling entry into the space, monitoring inside and outside the space during entry and this will put him into position to do this. If it is necessary to wear a SCBA during initial monitoring around the opening to the space, then the attendant can go off air once it is clear but keep the SCBA on or near in case of subsequent contaminant migration out of the space. This would allow him to continue his duties until the entry team has exited the space.
- Assessment of physical hazards.
  - ✓ The confined space entry permit or the energy control plan are good sources of information.
- Determine resource needs.
- Evaluate survivability profile (rescue or body recovery).

Phase III: Pre-Entry Operations

The pre-entry phase is where all preparations are made for entry into the space. Some of the tasks in the pre-entry phase will be worked on concurrently with assessment tasks. Evaluation of the information gathered during the assessment phase will determine the specific tasks that will need to be done.

Controlling Hazards

This includes the following two categories:

- Making the general area safe
- Making the rescue area safe

Making the General Area Safe

- Establish safety zones.
  - ✓ Create a marked area with tape or other means to establish working zones where only authorized personnel are permitted.
- Establish general area ventilation.
  - ✓ It may be necessary to set up some form of ventilation for the general area if a high level of contaminant migration is taking place.
- Assign an entry/exit point.
  - ✓ Decide where the entry/exit point for the operation will be so that it can be controlled and the rigging team can begin to set-up equipment as needed. If multiple points of entry are to be made, assure coordination.
- Eliminate all potential or actual ignition sources.
  - ✓ This may include hot work being performed in the area, vehicles, generators or other electrical equipment.

Making the Rescue Area Safe

Using the information from the hazard assessment, assign personnel to control the hazards present. Each of the tasks may take an individual or a team to complete them.

Rescue Group Supervisor

- Assign personnel to all positions needed to accomplish the tasks you have identified.
  - ✓ Always choose the personnel for the specific job at hand. Take into account the size of the space and the size of the rescuers.
- Perform lockout/tagout/blankout procedures.
  - ✓ Lockout/tagout needs to be done by someone familiar with the operation and should have been completed before the initial entry began. Off-site rescue teams should assign personnel to check the isolation of the space with a responsible person from the facility, placing rescue team locks and tags on each lockout point.
- Entry teams should always work in pairs.
  - ✓ No personnel should enter a confined space alone.
  - ✓ The buddy system is critical to successful operations.

- ✓ Exceptions may include:
  - ✓ When working in very small spaces, only a single entrant may fit.
  - ✓ A simple extrication such as clipping a retrieval line to the patient's harness.
- ✓ If the entrant will be out of visual contact with the attendant, a second entrant should make entry far enough to keep visual contact with the first entrant.

- For every entry team, there shall be a backup team.
  - ✓ The backup team shall be suited up and prepared to enter in the event of an emergency during the rescue.
- Each entry team should be assigned a unique call sign.
  - ✓ In the event that more than one team is working the space at the same time, each team needs a specific title.

Attendant

- Establish an entrant tracking system.
- Establish a communications system.
- Continuously monitor the atmosphere around the opening to warn of contaminant migration.

Ventilation

- Establish ventilation in the space.
  - ✓ The type of ventilation system will be determined by the hazards found during assessment.

Entry Team

- Entry team members should wear the PPE necessary for potential hazards as well as the present hazards.
  - ✓ Team members should help each other to suit up.
- The initial entry team should take an additional SAR unit for placement on the subject.
  - ✓ This can be attached to the primary entry teams air supply system.
  - ✓ An SCBA may be substituted for an SAR.

Backup Team

- Backup team members must be suited up with the same PPE as the entry team.

Rigging Team

- Identify some form of subject removal system.

  This could include the following:

  ✓ Tripods, davits or cranes.

  ✓ Rope or cable winch systems.

  ✓ Patient packaging devices.

  ✓ Teams assigned to work the removal systems.

  ✓ If working in a potentially flammable atmosphere, assure only non-sparking hardware is used for systems rigging.

Medical Group

- Set up for care and transportation as the patient is removed from the space.

Decon

- Set up for decontamination of entry teams, patient and equipment as they exit the space.

Logistics Section Chief

- Supervise air supply and medical unit.

Air Supply

- Set up primary and secondary air supply systems on-site and ready to become operational.

  ✓ One system for the entry team and patient.

  ✓ Separate system for the backup team.

- Establish a method to refill bottles, shuttle or move air to the primary site.

  ✓ This can be done by high pressure supply hose, cascade or the shuttle of 30- or 60-minute bottles.

- Assure adequate air line is on-site.

  ✓ Each section should be laid out and adequate air-line handlers and manifold operators assigned to each supply system.

- Assure that adequate accessories are present to perform emergency repairs to air systems.

  At a minimum this should include:

  ✓ Duct tape

  ✓ Teflon tape

- ✓ Toolbox
- ✓ Additional air fittings
- ✓ Extra hose

Medical Unit

- Set up rehabilitation and care area for entry teams.
- Perform pre-entry medical monitoring of entry and backup teams.
- Set up fluid rehydration for entry teams.

Pre-Entry Briefing

Prior to entry, each team shall receive a briefing from the rescue group Supervisor. This briefing is critical to the safety and success of any entry. The following minimum information shall be covered:

- Each team shall be advised of their expected task during the entry.
  - ✓ This should be specific: "Your job on this entry is to locate the subject, report any entrapment mechanisms and provide initial lifesaving care. Do not remove the subject unless it is quick and easy."
- Each team should be advised of emergency procedures within the space in the event of a rescue team problem.
  - ✓ Injury, loss of air, etc. can quickly create chaos. Knowing the emergency procedures should help all personnel to be aware of what the others are going to be doing, even if there is a loss of communications.
- Each team should be provided a site briefing.
  - ✓ This can be accomplished by showing the blueprints or maps located during the assessment phase or by generating your own map of the space. Remember that in some cases you will be entering a maze that is unfamiliar to you.
- During entry, the team should be aware of their surroundings and be ready to debrief when they exit the space.
- Each team should be advised of any time limits placed upon them.
- All personnel should know the hazards that may be encountered during entry, including information on the mode, signs or symptoms, and consequences of the exposure.

Once this information is relayed and acknowledged, the entry team can begin its entry operation.

Phase IV: Entry and Rescue Operations

Entry and rescue operations involve the actual placement of teams into the space, reconnaissance, location and removal of the subject(s) and entry personnel from within the space. No entry operation should begin until all of the pre-entry requirements have been completed. This is to assure the safety of the rescuers as well as the subject(s).

Responsibilities During the Entry and Rescue Operations Phase

Rescue Group Supervisor

- Coordinate all aspects of the entry, extrication and removal of the patient and entry personnel.

Attendant

- Continual atmospheric monitoring and recording of the measurements on a tactical sheet.
  - ✓ Evaluate monitor readings to assure that ventilation and safety are maintained.
- Logging of all entry times and names of entry personnel assigned to each team.
- Providing constant communications to the entry team and relaying status of the operation to the rescue group supervisor.

Ventilation

- Monitor the ventilation system to assure its continual operation.

Entry Team

- Work as a team and communicate planned actions to each other.
  - ✓ Coordinated action reduces time, stress and increases survivability.
- Assure adequate communications with the attendant.
  - ✓ If the attendant does not receive regular status reports when they are convenient to you, he will ask for them at the worst times.
- Watch your air lines and assist in the movement of them throughout the space.
  - ✓ Be careful not to crawl on, pinch or cut lines. Help each other and choreograph movements.

- Use entry/exit tag line.
  - ✓ This is a small line used to mark forward progress and the way out. A separate tag line is not necessary when using a SAR for the patient as the air line serves as a tag line.
- Beware of elevation differences and unstable footing.
  - ✓ Falls can kill in confined spaces.
- Beware of machinery, electrical and engulfment hazards.
  - ✓ Everything in the space should be considered hazardous, despite lockout/tagout procedures.
- Above all, remember the primary task and accomplish it.
  - ✓ Once you have located the subject(s), if patient removal can be accomplished quickly notify the attendant.

During the entry operations it may become necessary to perform multiple entries, using multiple teams. When this becomes necessary, as each entry team exits the space and before the next entry team enters, the following information should be exchanged between teams during a quick briefing:

- Location of patient(s) and their status.
- Any specific hazards which should be watched for.
- If the entry team has accomplished their primary task.
- An update of the map or blueprints as to the configuration of the space, patient location or other pertinent information.

Once the patient is located:

- Coordinate all movement with the rigging team and communicate progress or problems.
- Assure as much C-spine control as possible.
- Decide if the patient is to be removed head first or feet first.
- Beware of the use of wristlets on patients who have experienced burns. Skin may be pulled off during lifting operations.
- Watch the lower half of the patient's body during vertical hoist operations to assure it does not snag on objects protruding into the space.
- When moving patients through small openings, assure, whenever possible, that the rescuers are stationed on the egress side of the patient. Always try to avoid being blocked in by the patient during movement activities within the space. If this is not possible, make sure the following is communicated and agreed upon:

- ✓ Assure the rigging teams are aware of the problem and coordinate their activities accordingly.
- ✓ When the move is made, make it quickly and smoothly, leaving the space blocked for as little time as possible.
- ✓ Assure that all entry team and patient air lines are not pinched against the edge of the opening, cutting off air flow.

Backup Team

- Remain outside the space, suited up for immediate entry if the entry team needs rescue.

Rigging Team

- Assist the entry team into the space as necessary.
- Remain ready for an emergency evacuation of the entry team.
- Provide equipment requested by the entry team.
- Operate retrieval systems as necessary to evacuate the patient and entry team from the space.

Medical Group

- Have medical personnel staged and ready to receive the patient(s) for pre-hospital care.
  - ✓ If exposed to any hazardous substances that require an MSDS to be on file, the MSDS must be sent to the hospital with the patient.

Decon

- Decontaminate as necessary any personnel or equipment as it leaves the space.

Logistics Section Chief

- Coordinate logistics with the IC and operations section chief.

Air Supply

- Continually monitor breathing air supply.
  - ✓ Change air bottles as needed.
  - ✓ Request resupply of filled air bottles if needed.

Medical Unit

- Remain outside the space, ready to provide medical support to any rescue personnel.
- Perform medical monitoring when the entry team exits the space.
- Provide fluid rehydration for entry personnel.

Phase V: Termination

Once the patient(s) and the entry team have exited the space, assure their exit time is documented and that all personnel who were in the space are accounted for. All entry teams that leave the space should be debriefed for critical information discovered in the space. This may include:

- Location and position of patient(s).
- Condition in which the patient was found. Was he wearing his breathing apparatus? Was there evidence of a fall? This could become critical in an OSHA investigation or legal action.
- Have the entry team provide any additional information regarding mapping or drawing of the space for later documentation.
- Any specific problems encountered within the space regarding movement, hazards, air, etc.

All entry personnel shall be sent to rehabilitation for rehydration and medical monitoring. This information should be recorded. Consider critical incident stress debriefing for all personnel. Once all personnel needs are addressed, equipment needs should be taken care of. This should include:

- Inventory all equipment and ascertain if anything was left in the workspace.
- Account for all damaged equipment.
- Clean, perform maintenance, log, repack and restock equipment.
- Mark and/or tag any damaged or inoperable equipment.
- Have the contractor or responsible party seal the space and secure the entry points until completion of the accident investigation.

Incident commander, command staff and section chiefs should meet during the cleanup and discuss any problems encountered and make plans for a critique at the completion of operations or at a future date.

Typical Confined Space Equipment Set-Up

This typical confined space equipment set-up includes a tripod, ventilation, supplied air, communications and patient packaging.

Confined Space Entry Permit

CONFINED SPACE NUMBER	DATE OF ENTRY	DURATION OF ENTRY

PURPOSE OF ENTRY

ENTRANTS

Print Name ____ Signature ____ IN ____ OUT ____

Print Name ____ Signature ____ IN ____ OUT ____

ATTENDANT(S)

Print Name ____ Signature ____ IN ____ OUT ____

ENTRY SUPERVISOR(S)

Print Name ____ Signature ____ IN ____ OUT ____

ATMOSPHERIC MONITORING

ACCEPTABLE ENTRY CONDITIONS	TIME								
> 19.5% < 23.5%	OXYGEN								
< 10% LEL	FLAMMABLE								
PEL ______ ppm PEL ______ ppm									
CONTINUOUS MONITORING	TESTER'S INITIALS								

HAZARDS

☐ Mechanical ☐ Electrical ☐ Pneumatic ☐ Hydraulic ☐ Pipes or Ducts ☐ Engulfment

TEMPERATURE: ☐ High ☐ Low ☐ Hazardous Substance ____________

Other Permits Needed: ☐ Hot Work ☐ Safe Work

HAZARD CONTROL

☐ Ventilation Fans: Duct _____feet ☐ Saddle Vent with Bend MECHANICAL: ☐ Block Linkage ☐ Disconnect

ELECTRICAL: ☐ Lockout ☐ Tagout PIPES/DUCTS: ☐ Blind ☐ Disconnect

HYDRAULIC: ☐ Disconnect ☐ Lock Pump and Bleed PNEUMATIC: ☐ Disconnect ☐ Lock Compressor and Bleed

☐ ____________

PROTECTIVE EQUIPMENT

LIGHTING (EXPLOSION PROOF): ☐ Cord Light ☐ Caplamp ☐ Generator ☐ Power Cords: _____feet

☐ Fire Extinguisher ☐ Static Protection ☐ Warning Signs ☐ Cones ☐ Barricades/Tape

COMMUNICATIONS: ☐ Hardline ☐ Radios ☐ Hand Signs ☐ Rope Signals

☐ Tripod ☐ Davit ☐ Winch ☐ Retractable Lifeline with Winch ☐ Ropes ☐ Hardware ☐ Ladder

PERSONAL PROTECTIVE EQUIPMENT

☐ Body Harness ☐ PAL ☐ Safety Glasses ☐ Goggles ☐ Face Shield ☐ Hard Hat ☐ Ear Muffs/Plugs

☐ Coveralls ☐ Leather Gloves ☐ Splash Suit ☐ Chemical Gloves ☐ Overboots

RESPIRATORY PROTECTION

☐ SAR ☐ SCBA ☐ Cascade ☐ Airline: _____feet ☐ APR ☐ PAPR

CARTRIDGE: ☐ Organic Vapor ☐ Acid Gas ☐ Organic Vapor/Acid Gas ☐ HEPA ☐ Ammonia ☐ Dust/Mist

RESUCE AND EMERGENCY SERVICES

Rescue Team:	Phone No./Radio Call Sign:
Location of Phone:	Radio Frequency/Channel:

RESCUE EQUIPMENT: ☐ SKED ☐ Backboard ☐ Half Back

Were any unexpected hazards encountered during the entry? ☐ No ☐ Yes (explain below)

TIME PERMIT CANCELLED	PRINT NAME	SIGNATURE

A – 3

A full-size version of this form can be found in the appendix.

Sample Confined Space Rescue Permit

ASSESSMENT

DATE | LOCATION

TIME

RESPONSIBLE PARTY/CONTACT PERSON

NUMBER OF VICTIMS | TIME LAST SEEN | CONDITION

☐ ENTRY PERMIT AVAILABLE

DESCRIPTION OF SPACE | ACCESS

CONTENTS OF SPACE | ☐ MSDS AVAILABLE

HAZARDS IN SPACE

☐ Mechanical ☐ Electrical ☐ Pneumatic ☐ Hydraulic ☐ Other

ATMOSPHERIC: % Oxygen % LEL TOXICITY: % ppm of

Other Toxic Substances: % ppm of % ppm of

TIME TESTED | PERSON TESTING | METER CAL. DATE

☐ RESCUE ☐ RECOVERY (Acceptable Entry Conditions > 19.5% O_2 < 10% LEL/<PEL)

PRE-ENTRY

HAZARD CONTROL

VENTILATION: ☐ Positive Pressure ☐ Exhaust ☐ Local Exhaust ☐ Local Supply

MECHANICAL: ☐ Block Linkage ☐ Disconnect ☐ None ELECTRICAL: ☐ Lockout ☐ Tagout ☐ None

PNEUMATIC: ☐ Lockout ☐ Tagout ☐ None PIPING: ☐ Blind ☐ Disconnect ☐ None

HYDRAULIC: ☐ Lockout ☐ Tagout ☐ Bleed Lines ☐ Disconnect Lines ☐ None

EQUIPMENT REQUIRED

RESPIRATORY PROTECTION: ☐ SCBA ☐ SAR ______Ft. Airline

VENTILATION: ______Fans ______Ft. Duct ______Ft. Electrical Cord ☐ Generator

LIGHTING: ☐ Caplamp ☐ Hand Light ☐ Light Sticks ☐ Cord Light______Ft. Electrical Cord

(All equipment should be explosion-proof and equipped with GFCI)

ENTRY AND EXTRIFICATION: ☐ Tripod ☐ Davit ☐ Winch ☐ Rope M/A ☐ Belay Line ☐ Harness

VICTIM PACKAGING: ☐ Backboard ☐ Half Back ☐ SKED ☐ Litter ☐ Harness

COMMUNICATIONS: ☐ Visual ☐ Hardline ☐ Radio

ENTRY

Entry Team 1 | Phone No./Radio Call Sign:

Backup Team

Entry Team 2

Attendant

Atmospheric Monitoring Required:

☐ Continuosly Record on log every Min.

TERM.

Entry Terminated Time Date

Rescue Group/Entry Supervisor Print Signature

A full-size version of this form can be found in the appendix.

8 – 22

Sample Confined Space Tactical Log

DATE	INCIDENT
LOCATION	

TIME	ATMOSPHERIC MONITORING				ACTION TAKEN
	LOCATION	OXYGEN	% LEL	TOX	

A full-size version of this form can be found in the appendix.

CHAPTER NINE

RESCUE EQUIPMENT, KNOTS AND SYSTEMS

SCOPE:

This chapter serves as an introduction to rope rescue equipment, knots and systems.

TERMINAL LEARNING OBJECTIVE:

At the conclusion of this chapter, individuals will have knowledge of rope standards, equipment, knots, anchors, system safety factors and basic theoretical mechanical advantage.

ENABLING OBJECTIVE:

- Describe the standards and regulations that pertain to confined space rescue
- Describe the SI system and key terms used
- Describe the rope rescue equipment that may be used during a confined space rescue
- Describe a static system safety factor and what it means
- Demonstrate the ability to correctly tie the knots used in a confined space rescue
- Demonstrate the correct way to operate the Tandem Prusik belay
- Demonstrate the ability to correctly assemble and operate simple, compound and complex mechanical advantage systems

Equipment Standards & Regulations

Confined space entry is regulated by federal and, where superseded, by state law. Additional requirements come from the standards developed for rescue equipment and operations. By knowing and following the appropriate regulations and industry standards, the rescue will be conducted in the safest manner possible. This also will avoid regulatory citations, and if something should go wrong, mitigate possible litigation. Several of the most common sources of standards and regulations are the following organizations:

American National Standards Institute (ANSI)—Private sector standards coordinating center, U.S. & international standards. Consensus standards, voluntary compliance. Mandatory when referenced by OSHA in regulations.

ASTM International—Full consensus standards, voluntary compliance. Numerous standards in place concerning rescue equipment, testing, etc.

California Department of Industrial Relations Division of Occupational Safety & Health (Cal/OSHA)—Example of one of the many state agencies. Mandatory compliance, must meet or exceed federal regulations, fall protection and confined space covered, none on rope rescue specifically.

International Mountaineering and Climbing Federation—Formerly known as the Union of International Alpine Associations (UIAA). Develops European sport climbing and mountaineering equipment standards. Those voluntary compliance standards are often adopted by Comité Européen de Normalisation (CE), which establishes European equipment standards for many industries.

National Institute for Occupational Safety and Health (NIOSH)—The federal agency responsible for conducting research and making recommendations for the prevention of work-related injury and illness. NIOSH is part of the Centers for Disease Control and Prevention (CDC).

NFPA International—Industry consensus standards, voluntary compliance. Adopted by many fire departments.

Occupational Safety and Health Administration (OSHA)—Federal agency, mandatory compliance unless superseded by state law, fall protection and confined space covered, none on rope rescue specifically.

Units of Measure

The International System of Units (SI) is the global system of measurement. It is informally referred to as the metric system in the United States. Standards writing bodies, such as ASTM and NFPA, use this system when they establish minimum breaking strengths for rescue equipment. Because SI units are used on equipment, having a working knowledge of this system is important if the rescuer is to understand load ratings and safety factors of rescue equipment and systems.

SI is based on the decimal system (factors of 10) and uses a prefix to designate the order of magnitude of the base unit. For example the prefix *kilo* means 1000. 1000 Newtons = 1 kiloNewton (kN), 1000 grams = 1 kilogram (kg.). *Milli* means one-one thousandth. Millimeter (mm) = 1/1000 of a meter. *Centi* means one-one hundredth. Centimeter (cm) = 1/100 of a meter.

The following definitions relate to equipment and rescue system strength ratings.

Weight is a force. The weight of a body is a product of its mass and the acceleration due to gravity.

Mass is the quantity of matter in a physical body. It is a measure of a body's resistance to acceleration; mass does not change.

Pound (lb) is the US value for mass.

Kilogram (kg) is the SI metric value for mass (equals 1000 grams).

Pound-force (lbf) is the US value for force.

Newton (N) is the SI value for force. It is the force that when applied to a 1 kg mass gives an acceleration of 1 m/s^2.

Kilonewton (kN) is 1000 newtons. It is approximately 220 pounds.

Acceleration of gravity varies over the earth between 9.77 and 9.83 m/s^2. For the purpose of determining forces in rescue systems rounding up to 10 m/s^2 will simplify the process, while adding a small margin of safety to make up for errors that might be made in estimating the load.

Load can mean either mass or force, depending on its use. A load that produces a vertically downward force because of the influence of gravity may be expressed in mass units (100-kg load). Any other load is expressed in force units (such as 1 kN).

The following table provides the conversion factors from US units to SI metric units:

Conversion Factors from US Units to SI Metric Units

For This Quantity	To Convert From US Units	Multiply By	To Obtain SI Units	Multiply By	To Obtain US Units
Mass	pound-mass (lbm)	0.45359237	kilogram (kg)	2.204622	pound (lb)
Length	inch (in)	25.4 (exact)	millimeter (mm)	0.039370	inch (in)
	foot (ft)	0.3048	meter (m)	3.2808	foot (ft)
Speed	feet per second (ft/s)	0.3048	meter per second (m/s)	3.2808	feet per second (ft/s)
Acceleration	feet per second squared (ft/s^2)	0.3048	meter per second squared (m/s^2)	3.2808	feet per second squared (ft/s^2)
Force	pound-force (lbf)	4.448222 (4.4482216152605)	Newton (N) (kg • m/s^2)	0.2248	pound-force (lbf)
		0.0044482	kiloNewton (kN)	224.8089	

METRIC RULE OF THUMB:

A rule of thumb is an easily learned and easily applied procedure for estimating a value. For those of us not raised using the metric system, the following values have been rounded to simplify the process:

Mass = 2.2 lb = 1 kg

Acceleration of gravity = 10 m/s^2

(For exact conversion factors use the table above.)

To use the metric rule of thumb to determine the force placed on a rescue system assume that every person has a 100-kg mass (220 lb divided by 2.2 = 100 kg). Multiply that by the acceleration of gravity, 100 kg x 10 m/s^2 = 1000 N or 1 kN

One rescuer on the system = 1 kN

One rescuer and one subject on the system = 2 kN

Note: NFPA 1983 uses 1.33 kN (300 lb) for a Technical Use load and 2.67 kN (600 lb) for a General Use load.

Equipment Description and Capabilities

Rope

Rope is the primary tool used in many confined space rescues. Ropes differ in construction, material and size. These differences give ropes specific characteristics that must be considered when choosing the appropriate rope for a given task.

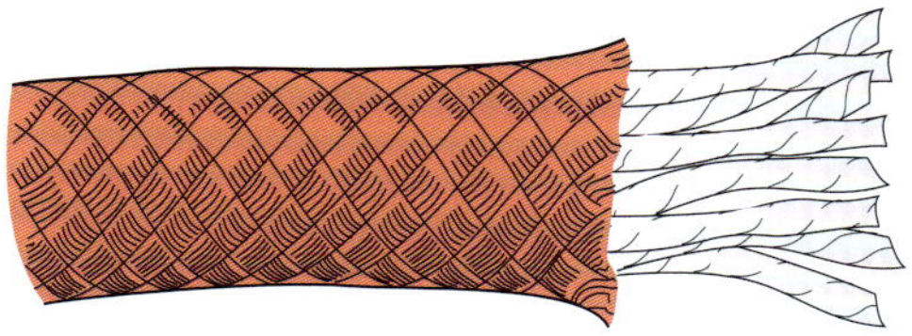

Kernmantle Rope

The rope most commonly used for rescue is of kernmantle construction. The word is German and literally means a core (kern) with a braided sheath (mantle). The main function of the mantle is to provide protection to the kern. The kern is the primary strength member and provides approximately 75% to 85% of the rope strength. Kernmantle ropes vary in construction and are typically made of nylon, polyester or a nylon/polyester mix.

Kernmantle rope is classified by the amount of elongation it exhibits at 10% of the minimum breaking strength.

Static rope is defined as a rope with an elongation of 1% to 6% at 10% of its minimum breaking strength.

Low-stretch rope is defined as a rope with an elongation greater than 6% and less than 10% at 10% of its minimum breaking strength.

Dynamic rope is a rope with an elongation greater than 10% at 10% of its minimum breaking strength.

100% polyester ropes are now finding wide acceptance in the rescue field. Although still meeting the specifications for static rope, these super static ropes, manufactured with High Tenacity Polyester (HTP) have as little as 1.5% elongation at 10% of its minimum breaking strength. These ropes are often used for highlines, deflections and any other situation where even a little stretch is counterproductive.

Static and low-stretch ropes are most commonly used in the rescue environment. Dynamic ropes are used for lead climbing and are designed to absorb the shock load of a falling climber by elongating. That elongation is essential for lead climbing but is not desirable in rescue. There are too many ledges, protrusions and other obstacles that a rescuer could hit should a fall occur. Even with their minimal elongation, static and low-stretch ropes still absorb some shock loading. Super static ropes are less forgiving. The best way to prevent injury to a falling rescuer is to ensure minimal drop distance by employing precise belay techniques.

The most common diameter for rescue rope is 12.5 mm (1/2 inch) with a minimum breaking strength of 40 kN (8,992 lbf). This rope has a working load of approximately 2.67 kN (600 lbf).

Rope Bags

Rope bags are nylon bags used to store ropes. They protect the rope from dirt and exposure, and when stuffed properly, they allow the rope to be deployed without tangling. It is helpful to have the rope bags color coded to identify the length of the rope stored inside.

Webbing

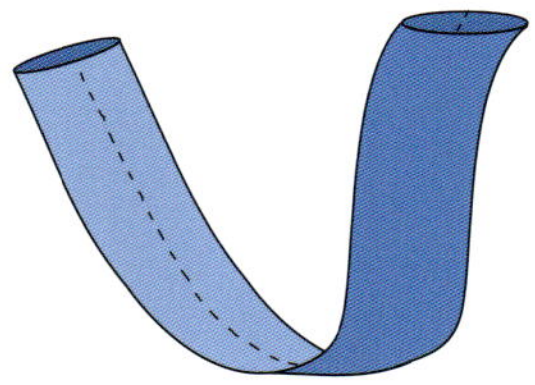

1-inch Tubular Webbing

Webbing is used for a variety of things, such as tying an anchor or lashing a patient into a litter. Webbing also varies in construction, material and size. The webbing most commonly used in rescue is 1-inch wide with either flat or tubular construction. The nominal strength of 1-inch tubular webbing is 4,000 pounds and is 6,000 pounds for 1-inch flat webbing.

Prusik Loop

A Prusik loop is most commonly made from 8-mm nylon Prusik cord. It is used to tie friction knots around the rope used in rescue systems. These friction knots are used as ratchets, as points of attachment in mechanical advantage systems and as belays for rappels and raising systems.

Anchor Straps

Sewn anchor straps have metal D-rings on each end. They are used to make quick, strong anchors for attaching ropes and systems. They are made of 1 3/4-inch flat nylon webbing. The minimum end-to-end breaking strength can be as high as 8,000 pounds but will vary between brands.

Load Release Strap

A Load Release Strap, a mariner's knot or a load-releasing hitch is used whenever a ratchet or Tandem Prusik is placed in a system. It allows a ratchet or Tandem Prusik under constant tension to be released without having to remove the load from the system.

Harness

Harnesses are made of flat nylon webbing that is sewn to create a safe, comfortable way to attach to a rope. After the harness is adjusted to fit, the webbing should be secured with a keeper or overhand knot. Harnesses that are used for fall protection and confined space rescue must meet ANSI standards as required by OSHA. They must be full-body (NFPA Class III) harnesses with attachment points that allow the wearer to be retrieved in the smallest possible profile. Prior to purchase, make sure the harness selected meets the practical requirements of confined space rescue as well as the regulations.

Carabiners

Carabiners are used to attach pieces of equipment together in rescue systems. Carabiners vary in size, material, shape, gate design and construction. Small aluminum non-locking carabiners are used in rock climbing. In rescue operations, only larger locking carabiners should be used. The larger carabiners are not only stronger, but accommodate multiple pieces of equipment. They also fit over larger diameter structures, such as the rail of a litter.

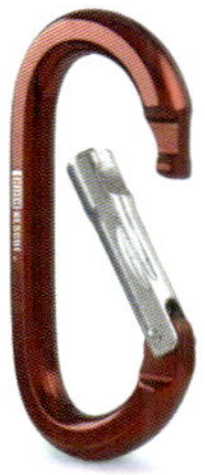

Non-Locking Carabiner

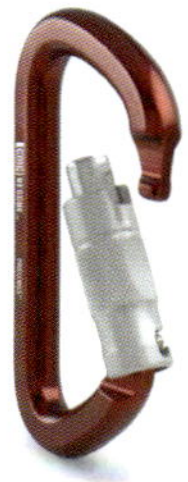

Locking Carabiner

The carabiner's shape also contributes to the strength. D-shaped carabiners are used, as opposed to oval or pear shaped. The D shape causes more of the load to be carried by the spine of the carabiner rather than the gate, thus increasing the effective strength.

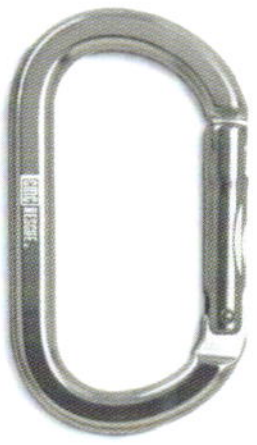

Oval

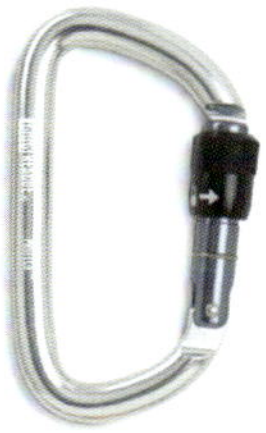

D

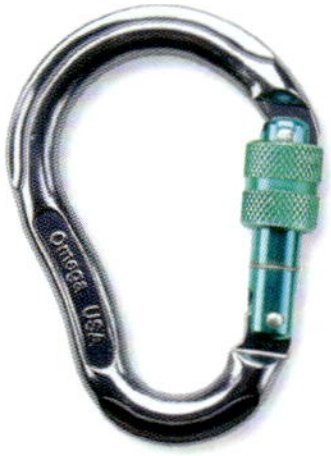

Pear

Because the spine of the carabiner is the strongest part, care should be taken to keep as much of the load as close to it as possible. As the load moves toward the gate the carabiner is weakened in a condition known as three-way loading.

Twisting forces on carabiners can side load the gate. That can cause the carabiner to fail well below its rated strength. With non-locking or unlocked carabiners, twisting can cause roll-out, resulting in the carabiners becoming separated. The chance of side loading carabiners can be increased by chaining or stacking carabiners to each other. That is especially true when the carabiners cannot move, such as when they are tight against a hard surface. Sometimes a second carabiner is beneficial because it can reduce the twisting motion in a system.

Screw Links

Screw links are often used in places where carabiners are not recommended, such as with anchor straps or other locations where three-way loading cannot be avoided. They are slower to open and close than a carabiner, but they can be more secure and they are less expensive.

Swivels

The addition of a swivel in the rope system can help prevent system lines from becoming tangled and wrapping around each other. They also allow the load to be easily rotated for better clearance through small openings.

Anchor Plates

An anchor plate is used as a collection point for attaching the pieces of equipment needed to build the different systems used in rescue operations. The holes allow each piece of equipment to have its own point of attachment. This makes systems less confusing and easier to manage.

Mechanical Ascenders

Mechanical ascenders, or rope grabs, are designed to hold a single person load while ascending up a fixed line. A complete knowledge of systems and potential forces that can be applied to them is necessary before using these pieces of equipment.

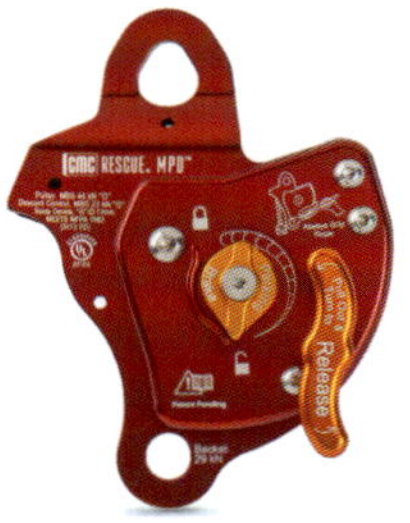

CMC MPD™

The MPD (Muti-Purpose Device) features a high-efficiency pulley, with an integral rope-grab mechanism, which allows it to be used as a lowering device on the main line and belay line systems and be quickly changed over to a raising system without switching out or replacing hardware. The combination of essential features into a single device simplifies on-scene rigging time. The device allows the main line and belay lines to be rigged for a twin-tension system. UL Classified to NFPA 1983 - General Use (13mm model) and Technical Use (11mm model) - Pulley, Descent Control and Belay Device.

Figure 8 Descenders

The Figure 8 descender is a friction device that is used for rappelling or lowering. When the rope passes over the many bends of the descender, the friction created gives the control needed to safely manage the descent. Because friction causes heat, the rappel or lowering speed must be controlled to minimize heat build up which could damage the rope.

The Figure 8 descender is small, light and easy to use. Once it is rigged, the rope cannot come out accidentally. The "ears" on the side of the Figure 8 allow it to be quickly and securely tied off. The ears were originally intended to prevent the rope from slipping over the top into a knot and stopping the descent at an inopportune moment.

A limitation of the Figure 8 descender is that the turns that the rope makes going through it cause twists in the rope, resulting in kinks that can cause rope management problems. The Figure 8 has a narrow range of friction adjustment and is best used for single person loads. It can be double wrapped for increased friction for pick-offs or lowering stretchers, but this increases the kinking.

Brake Bar Racks

Brake bar racks were developed by cavers for the long rappels into deep pits. They have the versatility to add or subtract friction during use and can control a rescue load easily. The rope runs straight through the rack, greatly reducing twisting, making it the preferred descender for a long lowering or rappel. It does take more practice to know how to use it well.

The rack's primary disadvantage, other than its weight, is the possibility of improperly attaching the rope. If the rope is woven on the wrong side of the bars, there is a high potential of popping the bars loose when the rope is loaded. This will result in the rope separating from the rack with a resulting fall. Special bars that act as indicators when combined with thorough familiarity with the equipment mitigates this risk.

CMC 3D™

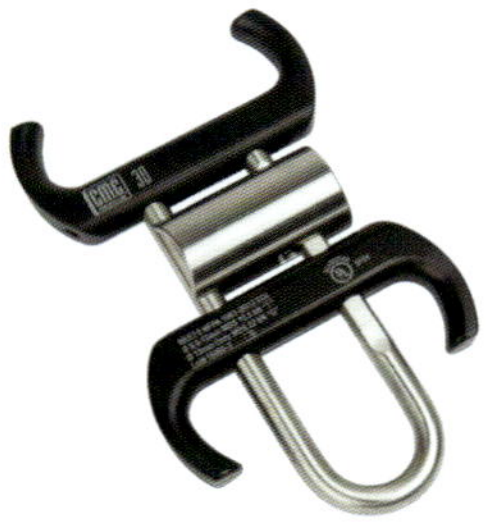

The 3D is a compact and lightweight locking-bar descender that features two sliding bars for friction control and efficiency. The horn design works well with 13mm rope and smaller, making it easy to add or subtract friction as needed throughout the evolution. The frame dimensions allow for larger rescue-sized carabiners to be used without any contact or friction with the rope. The descender can be loaded or unloaded while connected to your harness for rappelling, or connected to an anchor for one- or two-person rescue lowers. UL Classified to NFPA 1983 - General Use and Technical Use.

Edge Protection

Edge protection is used to protect ropes and anchors from being damaged on sharp or rough edges. Moving ropes should have edge rollers or pulleys to protect the rope from abrasion and to reduce friction. Fixed lines for anchors or rappels can be padded with anything that will soften the bends where the rope goes over the edge. Carpet, cardboard or rope bags will all work as edge pads.

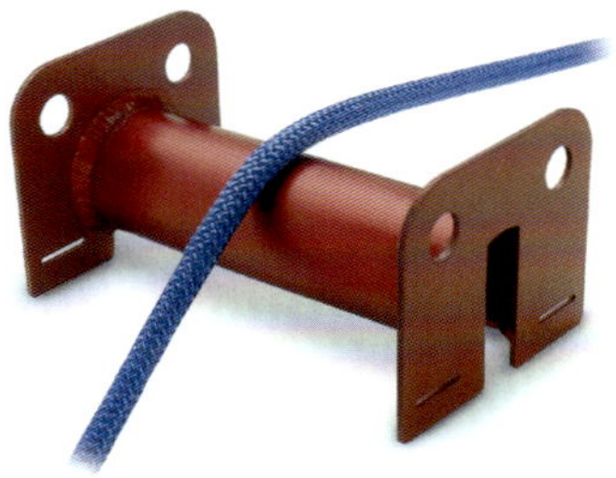

When working near an edge, all edge protection should be tied off to prevent it from falling. Use a length of cord or tubular web and tie it to an anchor.

Pulleys

Pulleys have two primary purposes in rescue systems. They are used to change the direction of the rope or to create mechanical advantage. A much rarer use is as a "traveler" in a high-line system. If the pulley does not move, such as when it is attached to an anchor, then it provides a change of direction for the moving rope. If the pulley is moving, it creates a mechanical advantage at the ratio of two to one. Explaining how pulleys do these different tasks can best be explained by relating them to levers.

How Levers Work

In the diagrams below, **R** is resistance (the load on the system), **E** is effort (the force required to move the load) and the fulcrum is the pivot point that determines the length of each lever arm.

First-Class Lever

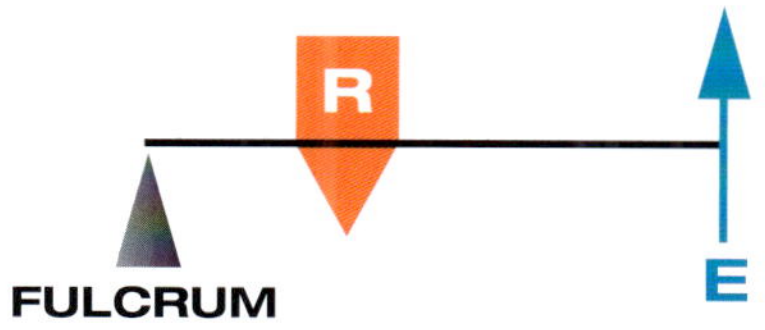

Second-Class Lever

How Pulleys Work

A **fixed pulley** can be looked at as a first class lever. The fulcrum is located at the pin in the center of the sheave, directly below the carabiner that anchors the pulley in place. Because the sheave is round, the distance from the pin (fulcrum) to the point the rope leaves the sheave and drops to the resistance (load) is equal to the distance from the pin to the point that the rope leaves the sheave and drops to the effort (force). The two lever arms are equal resulting in 1:1 mechanical advantage.

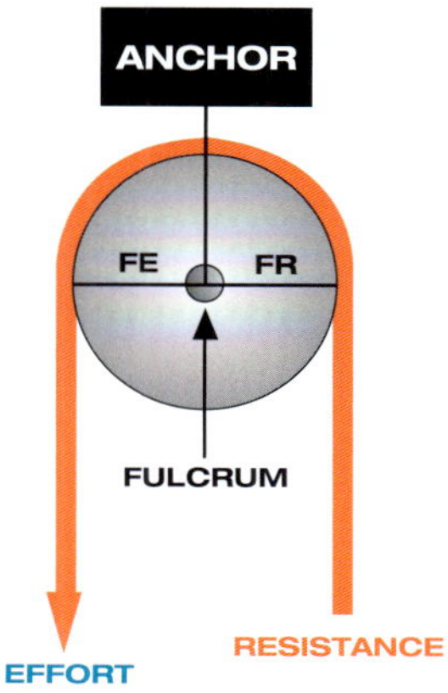

Fixed Pulley

A **moving pulley** can be looked at as a second class lever. The fulcrum is located on the edge of the sheave directly below the point the rope is attached to the anchor. One lever arm extends from the fulcrum to the pin, which is directly above the carabiner that attaches the resistance (load) to the pulley. The other lever arm extends from the fulcrum to the point where the rope leaves the sheave and goes to the effort (force). This lever arm is twice as long as the other arm, resulting in 2:1 mechanical advantage.

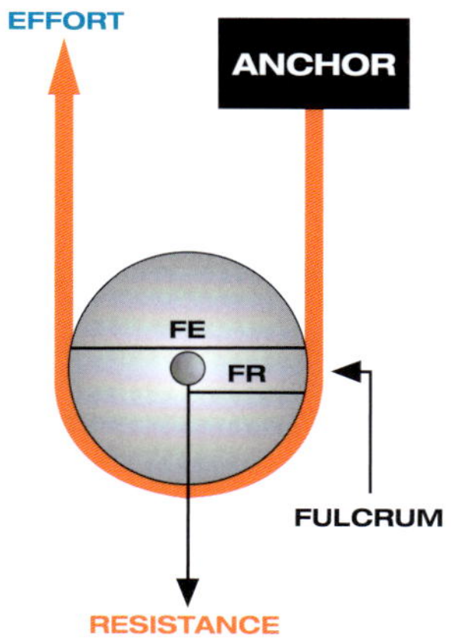

Moving Pulley

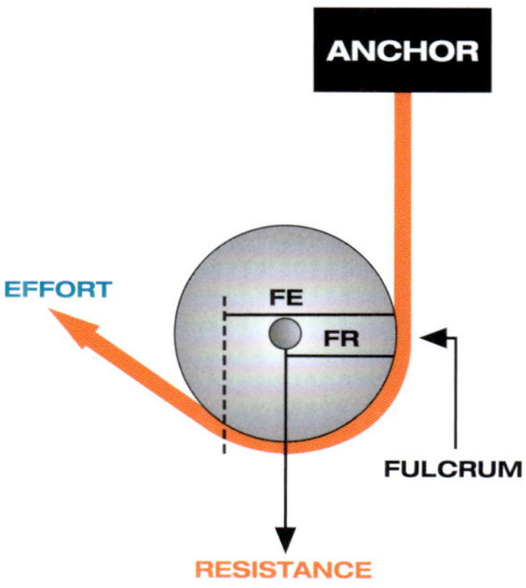

Moving Pulley
Pulled at Less Than 180°

When the rope is pulled through a moving pulley at less than 180°, the rope leaves the sheave at a point closer to the fulcrum, shortening the lever arm and reducing the mechanical advantage created by the pulley. This also decreases the mechanical efficiency by pulling the system sideways instead of in a straight line.

LSP Half-Back Extrication/Lift Harness

The Life Support Products Half-Back is a short cervical spine board incorporated into a harness. It has metal hardware so that it is rated for lifting and is easy to place on an unconscious patient. The aluminum spine board can be removed and the Half Back can be used as a harness by itself.

The patient hangs almost vertical and is lifted via rings on the front of the harness using a spreader bar.

LSP Miller Full-Body Splint

The Life Support Products Miller Full-Body Splint is a plastic, full backboard that has Velcro straps with which to attach the patient. It is narrower than many other backboards, which makes it preferred for work in small spaces.

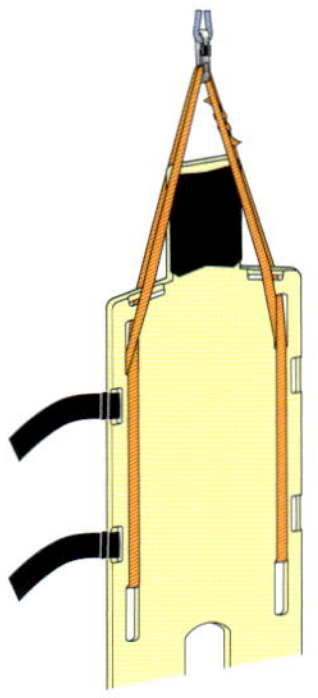

Lifting a Patient Vertically

Although the Miller Board was not designed for lifting a patient vertically, in many cases tight space and patient packaging requirements make it the best tool for the job. For vertical extrication, weave a 12-foot piece of 1-inch webbing through the hand slot at the waist, then up to the shoulder hand slot. Cross the web through and tie with a water knot. Do the same on the other side. The webbing should extend 12 inches above the top of the board. This set up spreads the load over more of the plastic structure of the should also be suspended independently in a harness.

Lifesaver Victim Chest Harness™

Another option is to use the CMC's Lifesaver Victim Harness in conjunction with the Chest Strap. Those can be wrapped around both the patient and the backboard, and the patient can be lifted vertically.

CMC/Skedco Drag-N-Lift Harness™

Because there is a need for a quick extrication device when spinal immobilization is not a concern, the CMC/Skedco Drag-N-Lift Harness™ was developed and engineered as a means to rapidly package and extract a patient from a confined space. The device enables the rescuer to provide a fast load-and-go technique that allows the patient to be packaged efficiently and securely for a vertical lift, horizontal carry out or drag extraction. A variety of handles and straps allow the patient to be transported by carrying horizontally, lifting vertically from a rope system, or pulled with the harness sliding on its smooth bottom surface. In the event that spinal immobilization is needed, the device is compatible with the Skedco Oregon Spine Splint (OSS II) and cervical collar.

Sked® Stretcher

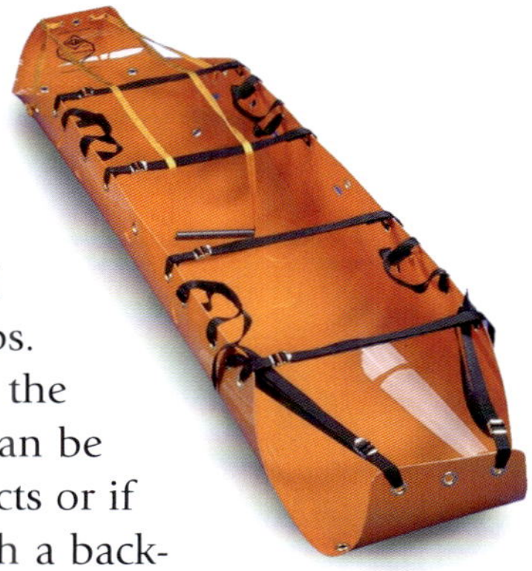

The Sked stretcher is the primary litter used in confined space rescue. Made from a flat sheet of plastic, it is carried tightly rolled up in a compact container. When laid flat, the sides are rolled up around the patient and held with the nylon straps. This stiffens the plastic and provides protection to the back and sides of the patient. The Sked stretcher can be used alone to slide a patient through pipes and ducts or if spinal precautions are required, it can be used with a backboard or the Oregon Spine Splint. A backboard may also be used to protect the patient if the Sked must be dragged across rough ground.

Half-Sked® Patient Drag

The Half-Sked is a drag sheet that is used in spaces with bends that are too tight for the full Sked stretcher. Because it is short it allows rescuers to bend the patient at the hip to turn corners. It features quick connecting side-release buckles but the patient should be secured in a harness for vertical lifting.

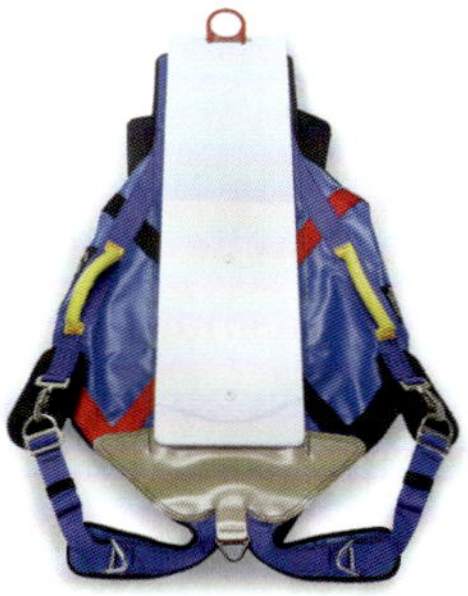

Spec Pak Patient Extrication System

The Spec Pak Patient Extrication System combines a short spine board with a full-body harness. It has a polyethylene backboard that provides both rigidity and a smooth sliding surface.

Wristlets

Wristlets are attached to a patient's wrists or ankles to allow them to be lifted or dragged through a very small opening. They should be used only with great care and as a last resort.

Equipment Inspection

All equipment should be carefully inspected when first received, whether it is new or used. Each time equipment is used, whether for training or rescue, it must be inspected and logged. When a problem is found, it must be identified and rectified, either by retiring the equipment or having it repaired by the manufacturer. Remember, your life, as well as the lives of your team members and the patient, depend on this equipment.

Rope

The following procedure can be used for inspecting rescue ropes. Hold the rope tightly in one hand and pull about 450 mm (18 inches) through, putting a slight tension on the rope. As you pull, use your hand to feel for soft spots or variations in diameter. Then twist the rope to visually inspect both sides. Continue the length of the rope. For additional information on rope inspection, see ASTM F1740-06 *Guide to Inspection of Nylon, Polyester, or Nylon/Polyester Blend, or both Kernmantle Rope.*

Webbing

Webbing is visually inspected in the same manner. Nylon is susceptible to damage from the ultraviolet rays of the sun and due to its large surface area, webbing is quickly affected. Fading and stiffness are signs of overexposure to the sun. Cuts and frays are best observed after washing, when the webbing is still wet.

Sewn Web Gear

The inspection technique is the same as for webbing. In addition, check the stitching for wear or pulled threads. Most manufacturers use contrasting color thread to make the stitching easier to see. Check for wear around all D-rings and buckles.

Carabiners

Visually check for distortion, cracks and wear. Make sure the gate will open and close all the way, that the spring still works and that the locking mechanism is working and free of dirt. The general rule is not to lubricate carabiners because the lubricant attracts and holds dirt in the hinge area, resulting in a stiff or stuck gate. One manufacturer, SMC, recommends the product LPS for lubricating carabiners. Another manufacturer, Omega Pacific, stated that they have been using WD-40™ for years without any problems. In any case, if you use a lubricant, use it sparingly.

Mechanical Ascenders

Visually check all parts of the ascenders. Check the cam for cracks, distortion and wear. Check the shell for sharp edges, distortion and elongation of the holes. Make sure the pin is not bent and that the indents still work. Check the cables, springs, cords and chains.

Pulleys

Visually check the side plates for distortion, cracks and elongation of the carabiner holes. Check the nuts for tightness, make sure the wheel turns freely and is free of dirt.

Descenders

Visually check for wear, cracks or distortion. Any nicks or sharp spots should be smoothed with emery paper. Aluminum descenders will readily show wear. When 15% of the material is worn through at any spot, the descender should be retired.

Equipment Repair

Equipment that is damaged in any way and appears to be repairable should be sent back to the manufacturer. Only the manufacturer has the correct parts and knows the type of materials that are needed to return the equipment to its designed strength.

Equipment Logs

Each piece of equipment that is used in rescue work should have a log maintained on its use, the recorded life history for that individual piece of gear. When repacking equipment after an operation, training or washing, inspect the equipment and log the findings and the usage. This will help determine whether it is still safe to use. Logs may offer some protection in the event of litigation by showing that the equipment was used, inspected and maintained properly. These logs should be kept where the equipment is stored and filled out immediately after each use. The log should document the following: the type of equipment, number, name of manufacturer, size, model, serial number, color, where and when it was purchased and the inservice date. After every use, fill in what it was used for, who inspected it and any comments on damage or wear.

Washing Equipment

All equipment should be kept clean and free of dirt and grit. Dirt can reduce the strength, effectiveness and life of rescue equipment. After gear has been washed remember to remark it if necessary and log that it has been washed.

Rope & Webbing

Washing rope and webbing in a washing machine is the easiest method. Run the machine through a cycle with plain water to rinse any harsh detergents from the machine before starting. Fill with cold water and put in the appropriate amount of detergent. We recommend a special product made for washing nylon rope, called LifeLine® Cleaner, which also works well on turnouts and hoses. Woolite® or other mild detergents that are safe for nylon may also be used. A muddy or especially dirty rope should be rinsed with water and possibly scrubbed with a nylon bristle brush before being put into the machine.

Double the rope (or webbing) and "daisy-chain" it. This keeps single lines from tangling or getting caught in the agitator. Put the rope in the machine and let it start to agitate. Turn the machine off and let the rope soak for 30 to 60 minutes, depending on how dirty it is. After soaking, run the machine through the gentle cycle. If the rope bag needs washing, put it in with the rope.

During the rinse cycle, add a small amount Downy® fabric softener to the load. (No more than one ounce of Downy to three gallons of water.) The fabric softener replaces the lubricant the rope loses during use and washing. This lubricant is put on the fibers by the nylon manufacturer to aid in rope construction. It lets the fibers slide along each other when the rope is loaded and unloaded. Without the lubricant, the fibers tend to bind together, reducing the strength of the rope.

Air dry the rope and webbing in a cool, shaded place. Do **NOT** dry nylon products in the sun because of the damaging effects of ultraviolet rays to nylon. If necessary, the ropes can be stuffed into the bags wet. The ropes may mildew but this does not adversely effect the rope.

Sewn Nylon Gear

The D-rings and buckles tend to damage washing machines. Use the LifeLine Cleaner and soak them in a tub. Light scrubbing with a brush should remove heavy dirt deposits. Rinse and air dry the same as rope and webbing.

Hardware

When necessary, wash hardware in warm, soapy water. Be sure to get all of the dirt out of any moving parts. Rinse and towel or blow dry. As we mentioned under equipment inspection, the rule against lubricating carabiners and hardware appears to be changing. Use of a dry or non-sticky lubricant following washing may help preserve the life and performance of your hardware.

Retiring Equipment

Determining when to retire expensive equipment can sometimes be difficult. When there are obvious cracks, tears or breaks, it is easy to put it out of service. When it is not so obvious, no visible damage but a possible shock load, or a slight irregularity in the core of a rope, it is best to remember that your life might be the next one to hang on that system. The only way to test the strength is to test it to destruction, which is equal to retirement. **WHEN YOU ARE NOT SURE, RETIRE IT**.

When it is retired, make sure it can't accidentally end up back in a rescue system. A rope that ends up in the back of some truck to tie equipment down, or carabiners that end up holding keys on somebody's belt loop, might be recognized as rescue equipment and returned to service. Cut it up, paint it black, anything that **PERMANENTLY** marks it as damaged equipment.

Note: Black color code designates training equipment for some organizations.

Rope Life

ASTM and NFPA put the maximum service life of life safety rope at 10 years from date of manufacture.

Knots

Although there are many knots that may be used for the same purpose, selecting a small number of knots and knowing them well is the best policy. Knowing what knot to look for in a given position in the system makes checking the system easier.

The tensile strength of a rope is derived from a test that pulls a straight rope with no bends until it breaks. A rope is weakened by making bends; the sharper the bend the more strength is lost. Since a knot is a series of bends, we pick knots that have the largest possible bends while still being manageable. This gives them the highest strength possible. Knots can be rated for strength by the percentage of rope strength that remains when a knot is tied in the rope. This is called **knot efficiency**. For example, a rope with a 100-pound tensile strength that breaks at an 80-pound load with a particular knot in it means that this knot has a 80% efficiency rating.

It is important to use knots that are simple to tie and then easy to inspect to assure that they are tied correctly. Many of the knots used for tying rescue rope are from the figure-eight family of knots. They are similar to tie and look alike when tied correctly. Knots in web are based on the overhand family of knots.

The ability to untie knots after they have been loaded is also important. Many ropes would be wasted if the knots could not be untied after they were used.

The knots that are included in this manual were picked because of their strength, simplicity and ease of untying after having been loaded. The common knots sometimes have several names. The names most familiar to the rescue community are used in this manual.

After careful consideration, we no longer emphasize the need to add a safety knot to the knots shown in this manual. One of the reasons why we have selected the knots shown is that they are inherently safe and do not come loose when the tension of a load is removed from them. We continue to stress the importance of a safety check to be sure the knot is tied correctly and the need to pull the knot snug before it is loaded. There should be at least six inches (15 cm) of "tail" remaining from knots tied with rope and three inches (7.5 cm) remaining from knots tied with webbing after they are pulled snug.

Knot Efficiency in 1/2" Kernmantle Life Safety Rope

	100% Nylon Sheath/Core	100% Polyester Sheath/Core
Alpine Butterfly - Rope pulled end to end	69%	60%
Alpine Butterfly - Rope pulled loop to end	80%	61%
Bowline	71%	58%
Double Loop Figure 8	77%	67%
Double Fisherman's Bend	76%	80%
Figure 8 Bend	60%	66%
Figure 8 on a Bight	78%	70%
Scaffold Knot	78%	69%
In-Line Figure 8 - Rope pulled loop to end	54%	49%
In-Line Figure 8 - Rope pulled loop to end	77%	63%

OVERHAND KNOT

Used to tie off knots in webbing and in making other knots.

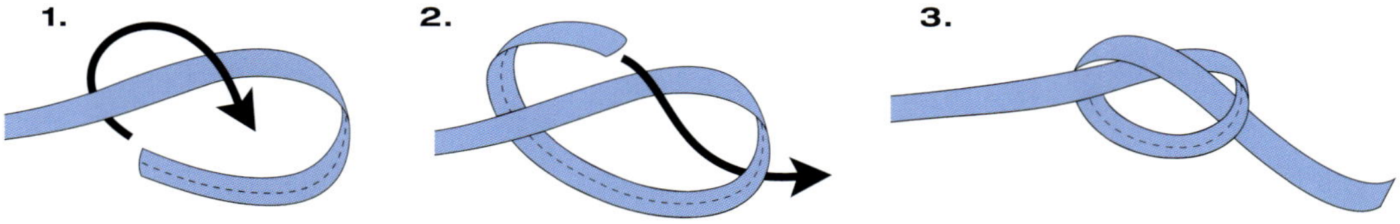

WATER KNOT (OR RING BEND)

Used to tie webbing together, also called a Ring Bend or Overhand Bend.
Knot efficiency = 64%.

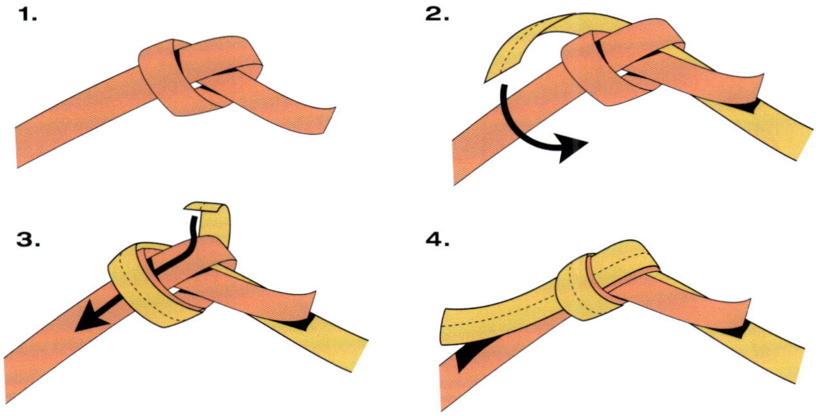

Note: The Water Knot should have 3 inches (8 cm) of "tail" remaining from the knot when it is tied and pulled snug.

ROUND TURN AND TWO HALF HITCHES

Used to tie off webbing.

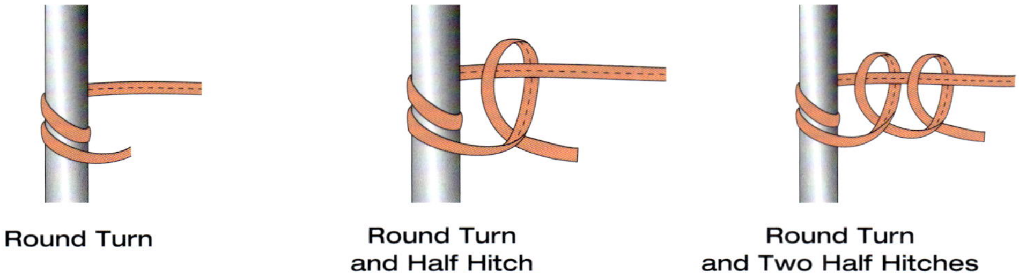

MODIFIED TRUCKER'S HITCH

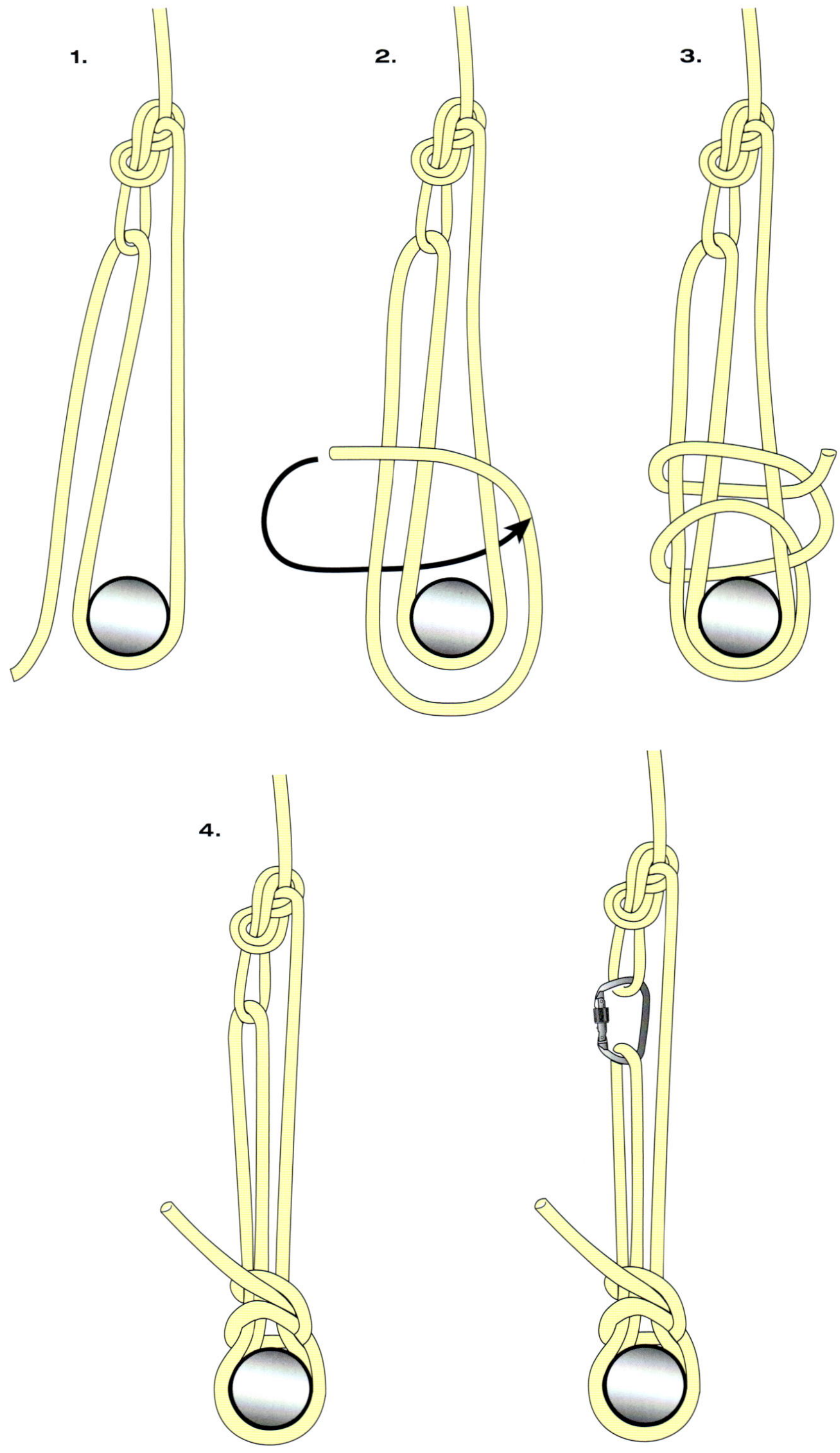

Alternate Method
Using Carabiner

DOUBLE FISHERMAN'S BEND

Used to tie Prusik loops.

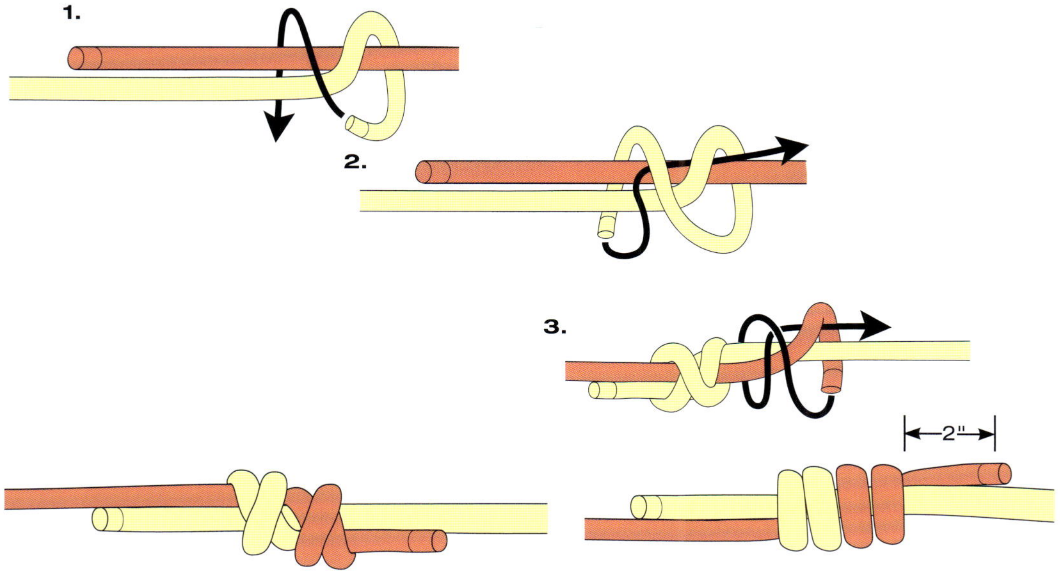

Front Back

Note: The Double Fisherman's Knot used to create a Prusik loop should have 2 inches (5 cm) of "tail" remaining from the knot when it is tied. When the knot is loaded and tightens, some of that excess will be taken in.

PRUSIK HITCH

Friction knot

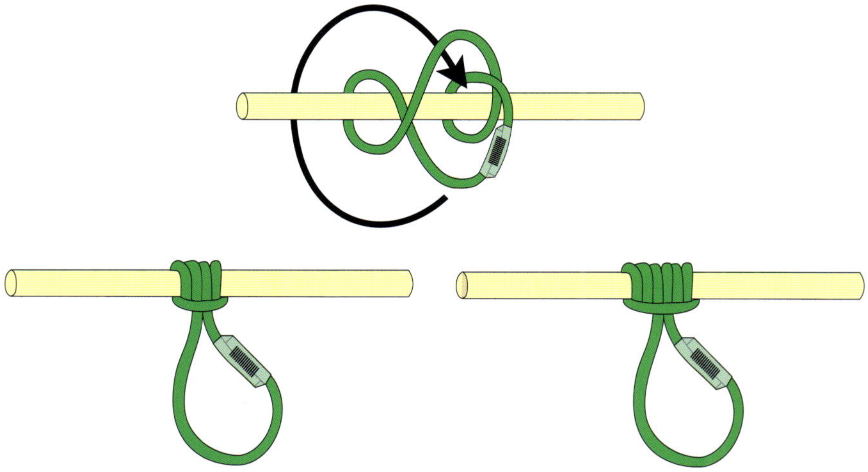

Two-Wrap Prusik Three-Wrap Prusik

FIGURE 8 KNOT

Used to tie other knots and as a stopper knot.

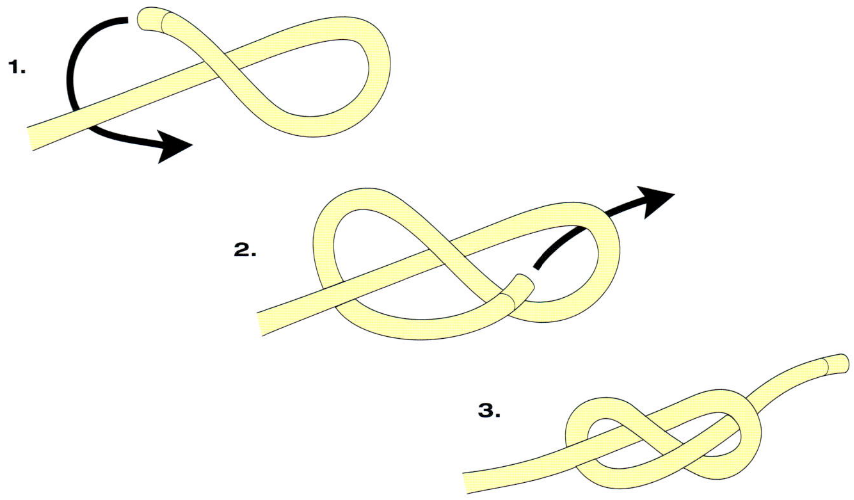

FIGURE 8 BEND (FIGURE 8 FOLLOW-THROUGH)

Used to tie two ropes together. .

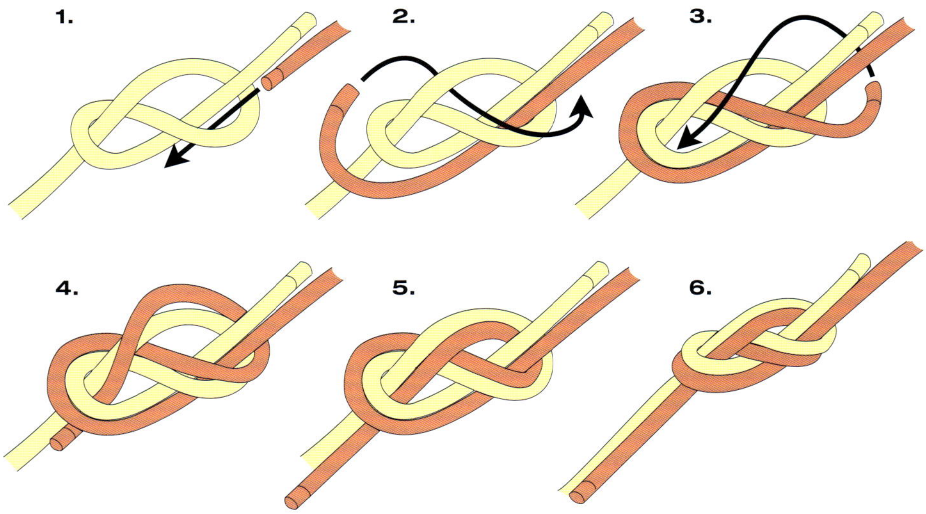

Note: The Figure 8 knots shown on this page should have 6 inches (15 cm) of "tail" remaining after the knot is tied and pulled snug.

FIGURE 8 ON A BIGHT

Used to make a loop in a rope.

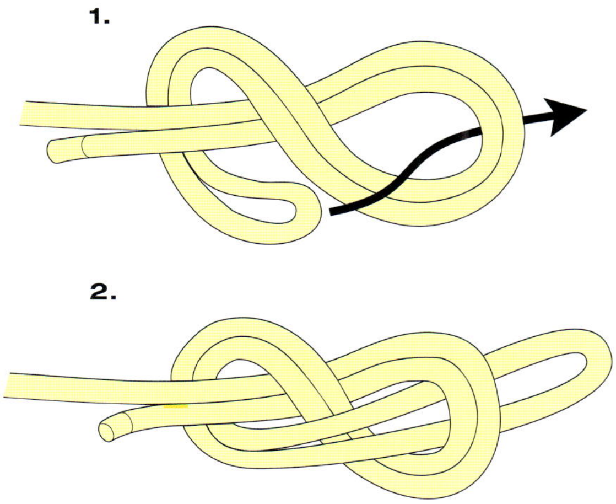

Note: The Figure 8 on a Bight should have 6 inches (15 cm) of "tail" remaining after the knot is tied and pulled snug.

FIGURE 8 FOLLOW-THROUGH LOOP

Used to tie a rope around an anchor.

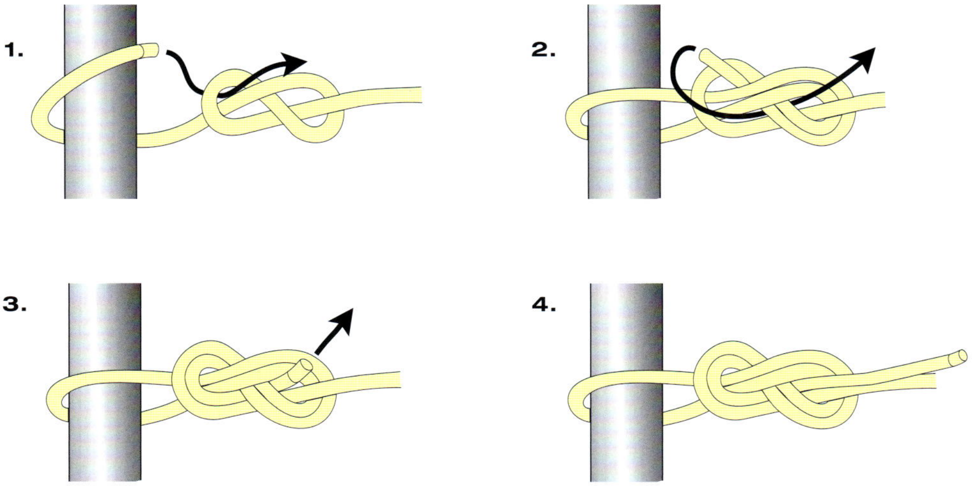

Note: The Figure 8 Follow-through Loop should have 6 inches (15 cm) of "tail" remaining after the knot is tied and pulled snug.

DOUBLE-LOOP FIGURE 8

Used to tie a double loop in a rope.

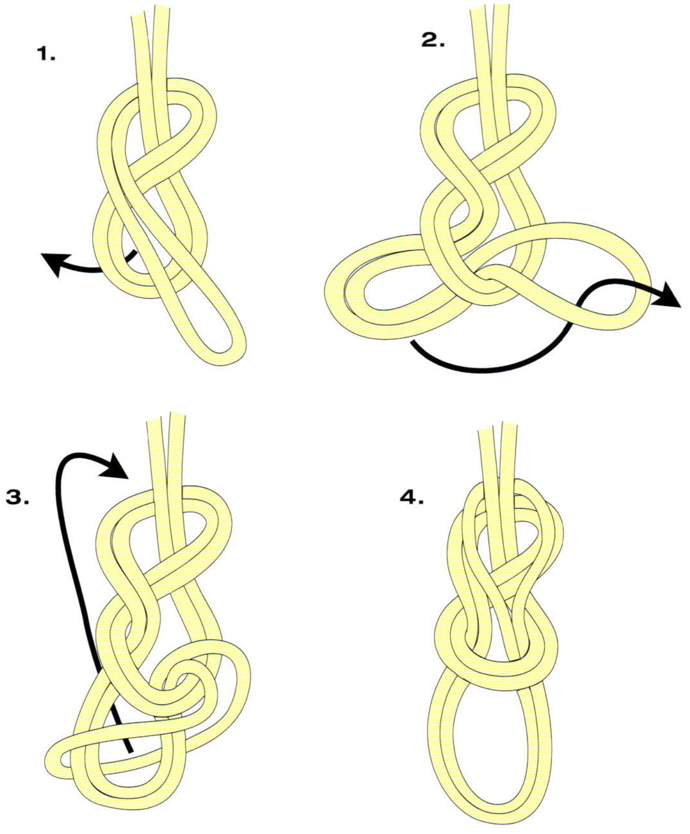

HIGH-STRENGTH TIE-OFF

Used to tie rope to an anchor. Knot efficiency = 100% (if the diameter of the anchor is at least 4 in).

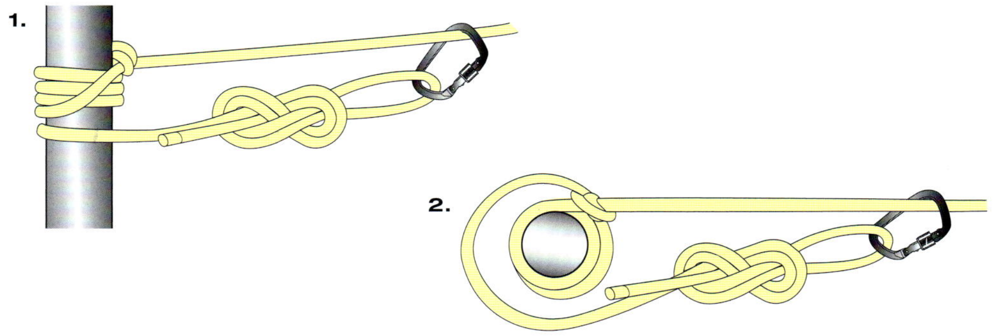

TENSIONLESS HITCH

A very quick anchor that eliminates the knot at the anchor point, if you have enough wraps around a large, not-too-smooth surface. Knot efficiency - 100% (if the diameter of the anchor is at least 4 in).

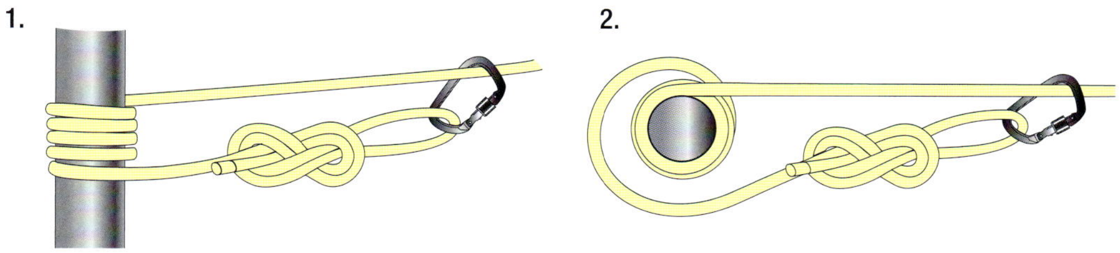

HANDCUFF KNOT

One method to secure a litter patient's hands.

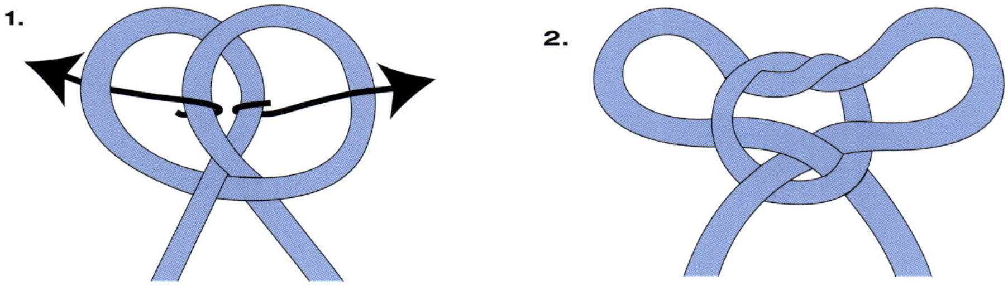

HASTY CHEST HARNESS

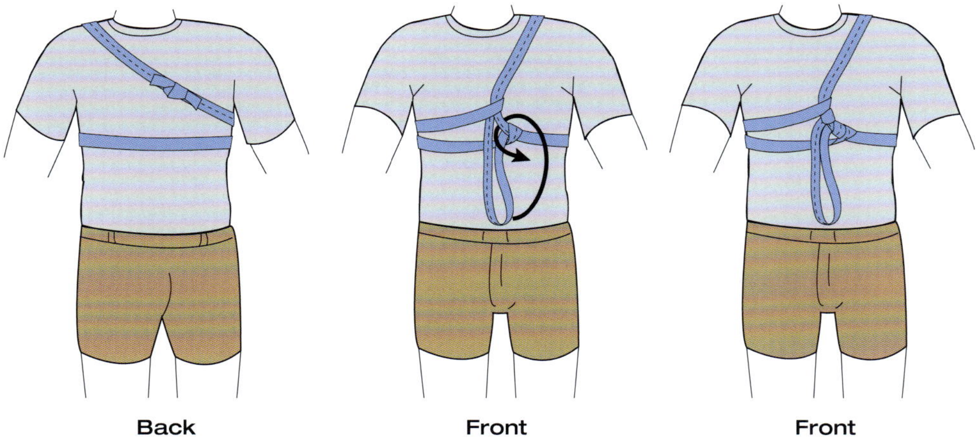

Patient Packaging

CHEST HARNESS

OPTION A

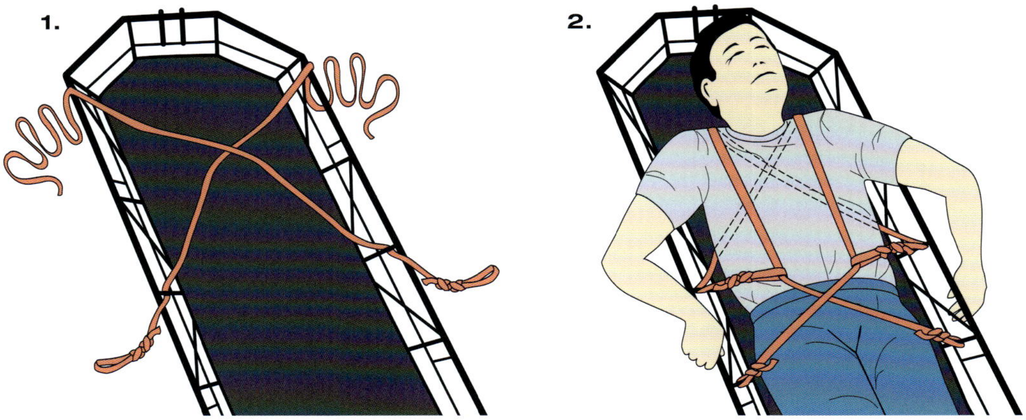

OPTION B

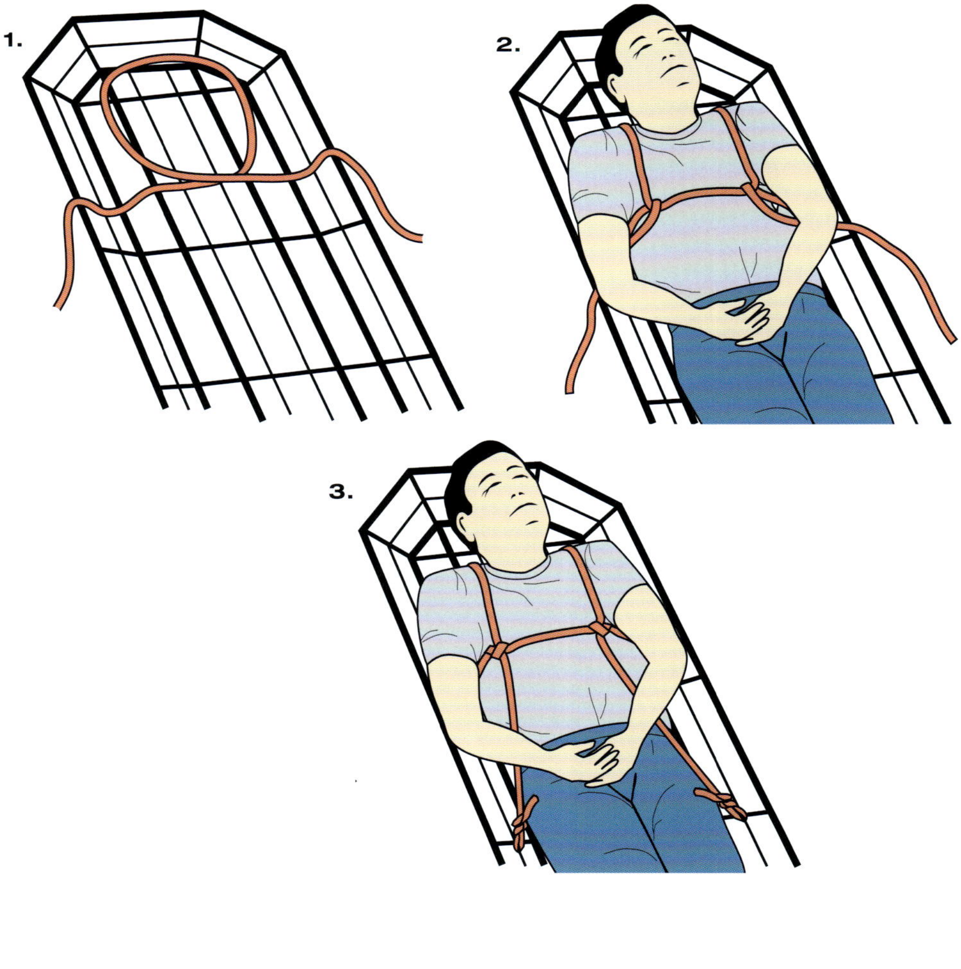

PELVIC HARNESS

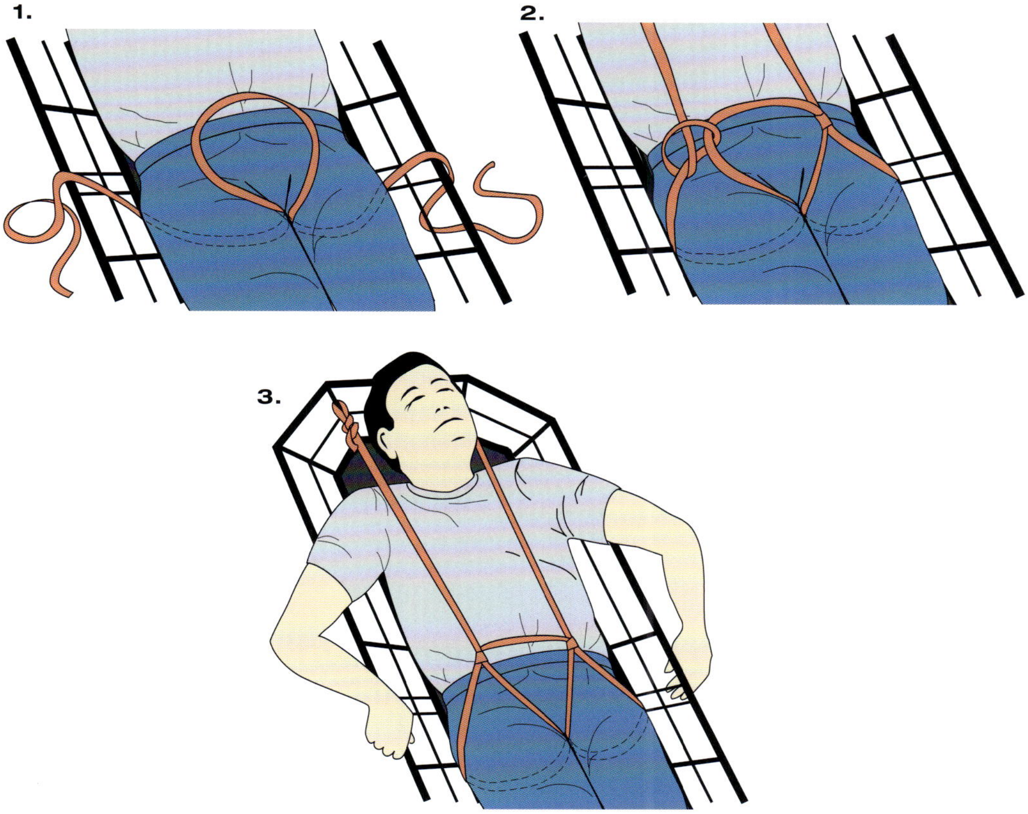

LIFESAVER VICTIM HARNESS

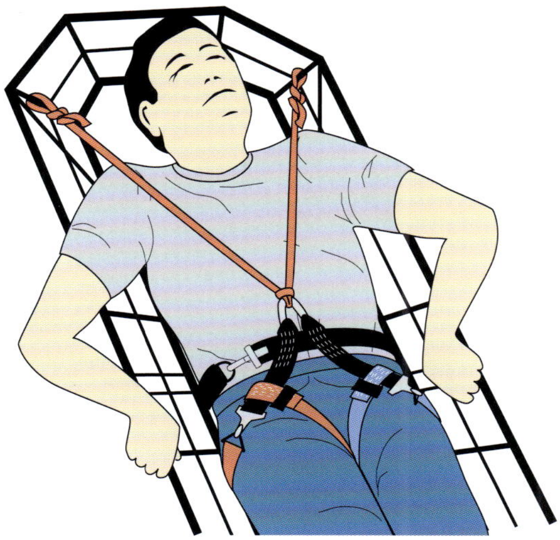

EXTERNAL LASHING

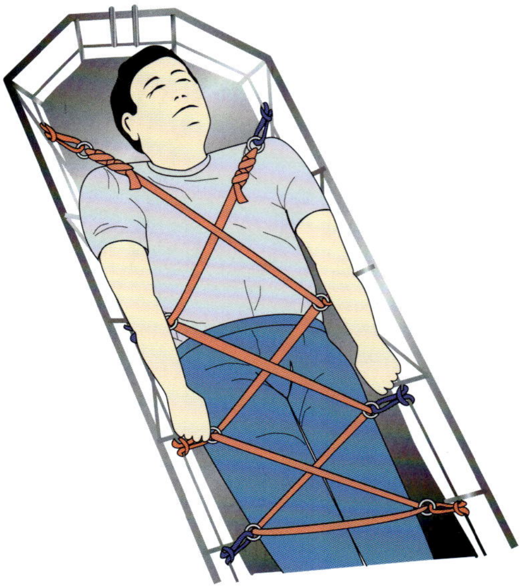

HOW TO TIE WRISTLETS AND ANKLETS OUT OF WEBBING

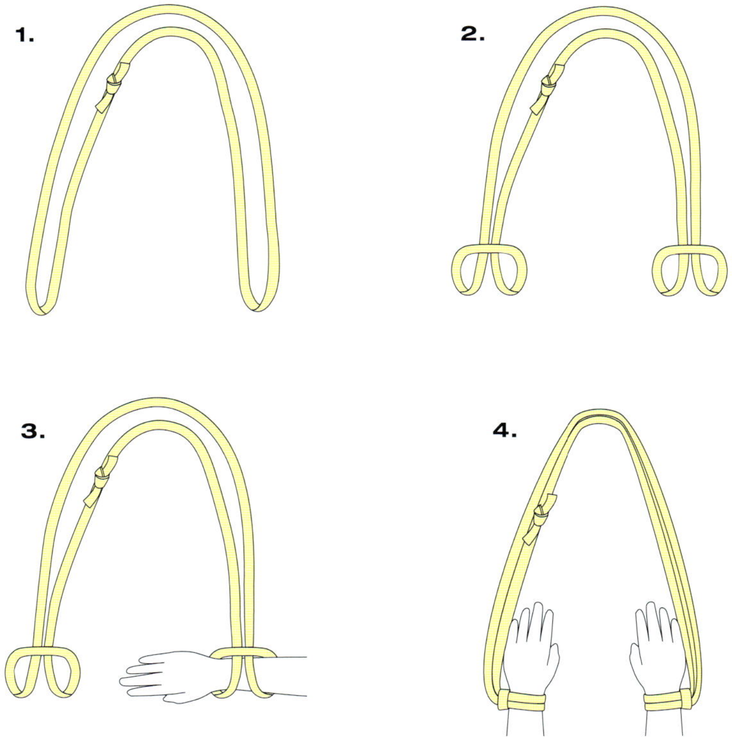

Anchors

Anchors are the foundations that all rope systems are built on, and placing a good one is a combination of technology and art. It does not matter if the rope will hold 7,000 pounds if the anchor will only hold 70 pounds. There is no rule of thumb that will substitute for experience and judgment. Whether you are working on cliffs or in industrial settings, the best way to learn anchors is from hands-on practice working with an experienced rescuer.

The importance of a strong anchor should be obvious, but how or where to attach the anchor can affect the strength. Tying to the base of a tall anchor reduces the leverage on the anchor, making it as strong as it can be. Likewise, picking the I-beam that is held on with 12 bolts instead of the I-beam held on with four bolts would be best.

The location of the anchor can make a difference between a system that works and one that does not. If the anchor is not in line with the load, or if the anchor is too close to the edge, a change of direction can be employed in the system to turn the rope, so it lines up with the load or can extend the rope in another direction, making up for the lack of room at the edge.

THE EFFECT OF LEVERAGE ON AN ANCHOR

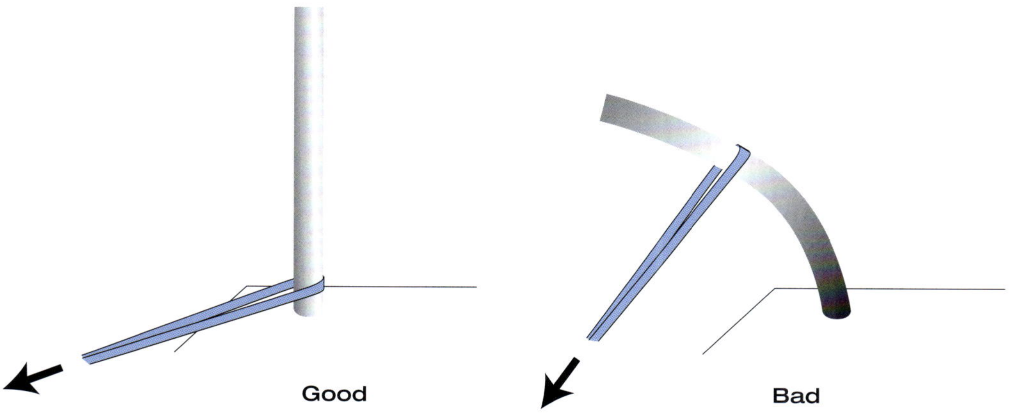

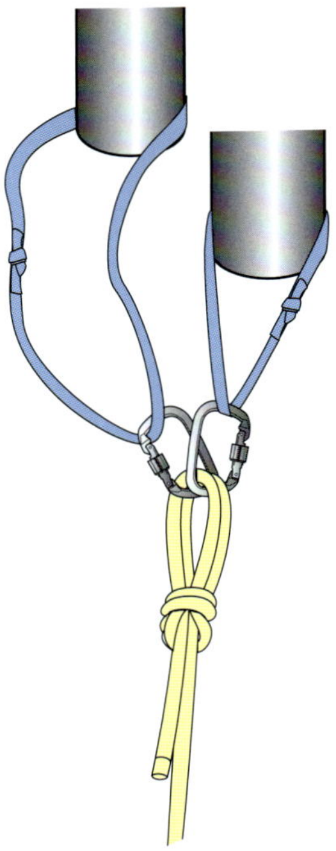

Backed-Up Anchor

What the anchor is tied to is important. Is it wrapped around a steam line? Will the heat effect the nylon? What if a sturdy pipe is located near an acid or other chemical that degrades nylon? Are there sharp edges that can cut through the straps? These problems are addressed by adding padding to protect the straps from sharp edges and maybe even the heat, or selecting a better anchor somewhere else. Nylon equipment used near a chemical that may degrade it must be retired immediately.

Many times the anchor is out of sight and there will be no warning if it starts to fail. For this reason, it is common practice to back up the anchor with another anchor of equal strength. Use two separate straps and try to locate the two anchors in line to avoid a pendulum should one anchor fail. A backed-up anchor is considered bombproof. If your primary anchor point is so strong that the possibility of anchor failure is beyond doubt, both straps can be tied to that one anchor point. Because we have still backed up the hardware and the straps, this would be a bombproof anchor also.

The length of the straps is also important. Wrapping the straps too tightly around the anchor increases the load on the straps. Watch that the interior angle of the straps stays smaller than 90°.

Another problem can arise when load-sharing anchors are constructed. These anchors are useful when the system needs to be centered between two anchors. The problem becomes evident when there is a load shift. The load that was shared equally by both anchors is now transferred onto one anchor. The solution is to build a load distributing anchor system.

ANCHOR FORCE AND LOAD RELATIONSHIPS

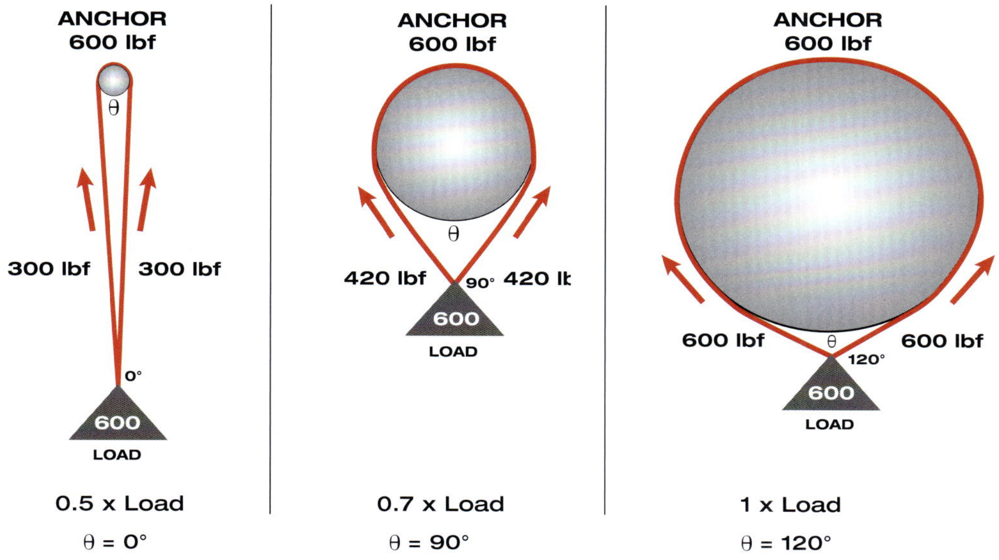

Figure 8-2a: Forces on Anchor Web (Rope) Due to Critical Angle (θ)

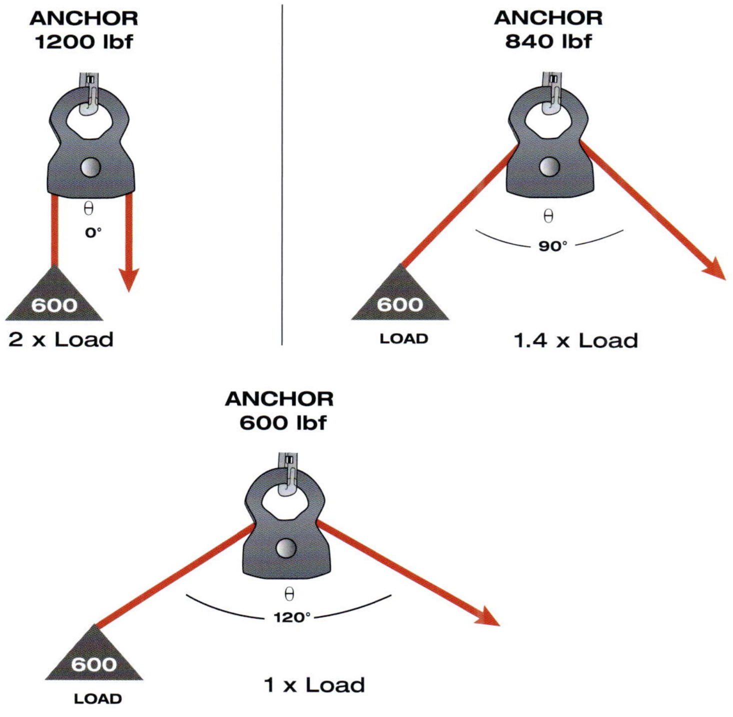

Figure 8-2b: Forces on a Change-of-Direction Anchor as a Function of the Critical Angle (θ)

One-Inch Webbing Anchors: Minimum Breaking Strength of Common Configurations

		Tubular Web lbf. (kN)	Flat Web lbf. (kN)
Web Strength	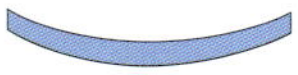	4,340 (19.31)	6,000 (26.00)
Choker or Girth Hitch	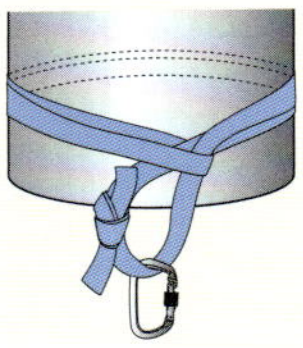	4,799 (21.35)	8,776 (39.04)
Single Loop (90° Internal Angle)	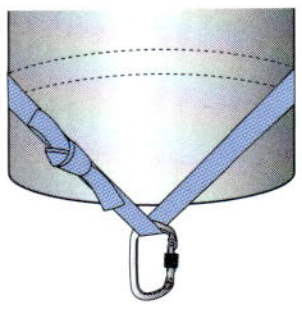	4,832 (21.50)	6,130 (27.27)
Wrap 2, Pull 1 (90° Internal Angle)	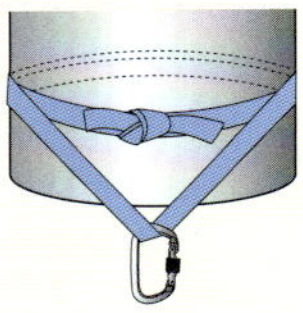	5,510 (24.51)	8,098 (36.02)
Redundant Double Loop (90° Internal Angle)	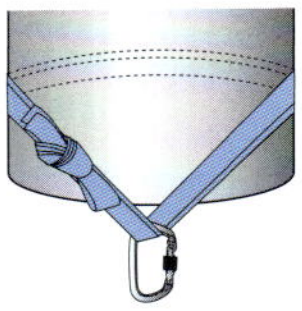	7,777 (34.59)	10,786 (47.98)
Wrap 3, Pull 2 (90° Internal Angle)	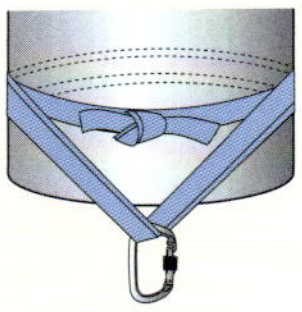	7,899 (35.14)	10,507 (46.74)
Basket (90° Internal Angle)	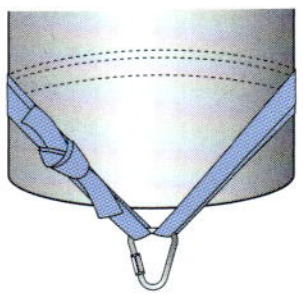	8,464 (37.65)	12,989 (57.78)
Double Loop (90° Internal Angle)	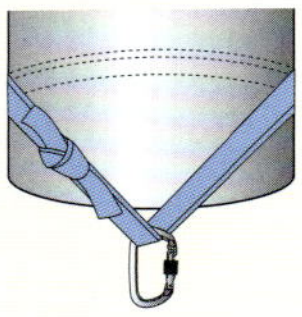	8,716 (38.77)	10,538 (46.88)
Redundant Wrap 2, Pull 1 (90° Internal Angle)	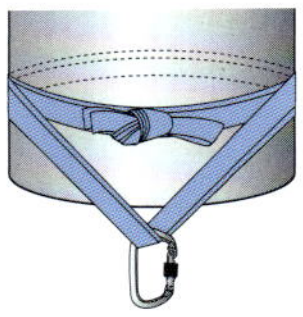	9,700 (43.15)	11,458 (50.97)

Load-Distributing Anchors

A load-distributing anchor centers the system between two or more anchor points and solves the problem caused by a load shift as the system is loaded. The load distributing anchor allows the load to be distributed to each anchor point by permitting the point of attachment to shift within the anchor. This distributes the load to each point in the system. Once the full load is on the system, the friction is too great to allow further distribution. The ability to shift the point of attachment within the anchor creates a possible problem. If one of the anchor points fails, the shift to the remaining anchor points will cause a drop in the system. If the anchor legs are long, this drop can create a dynamic-loading situation that may cause a failure of other anchor points. Keeping the anchor legs as short as possible, a maximum of 12 inches, reduces this possibility. This system makes it possible to combine a number of anchors that may not hold the load individually into a anchor system that will hold the load by distributing the load between all of them.

LOAD-DISTRIBUTING ANCHOR SYSTEMS

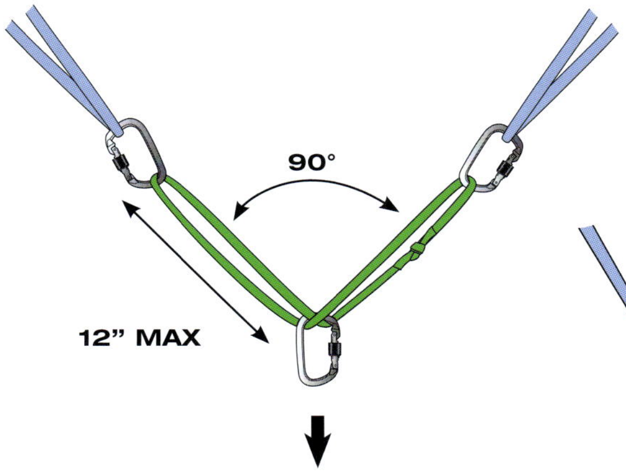

Two-Point
Anchor System

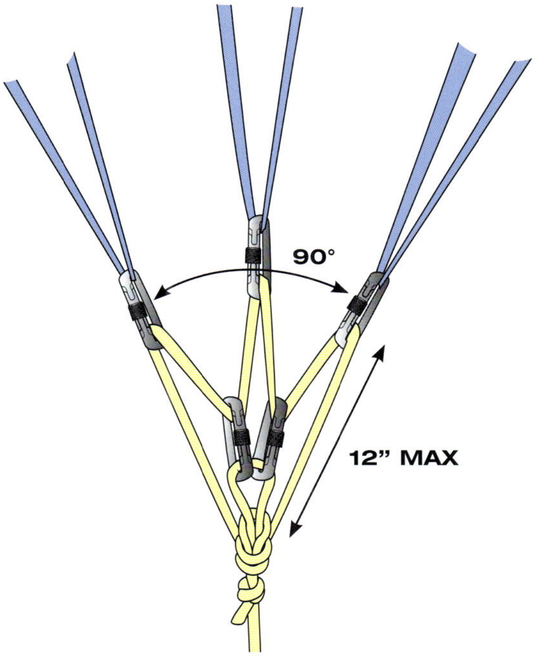

Multipoint
Anchor System

Rescue Systems

Rigging a rescue system starts with a bombproof anchor that will safely hold the anticipated load. Next, hardware and rope are added, completing the system. Being prepared to modify the system during the rescue allows the rescuers to deal with unexpected changes to the rescue plan. Having a standard complement of equipment that is set-up every time a system is rigged allows immediate modification.

An anchor plate works well to keep system equipment organized. Use the individual holes to attach the various pieces of hardware that may be needed during the course of the operation. Pre-rigging the anchor plate with the equipment necessary to build a lowering system and a mechanical advantage raising system and then storing it in a bag ready to go speeds the set-up for all operations. Some people refer to an anchor plate with attached hardware, as shown at right, as an RPM (rack/pulley/mariner's hitch).

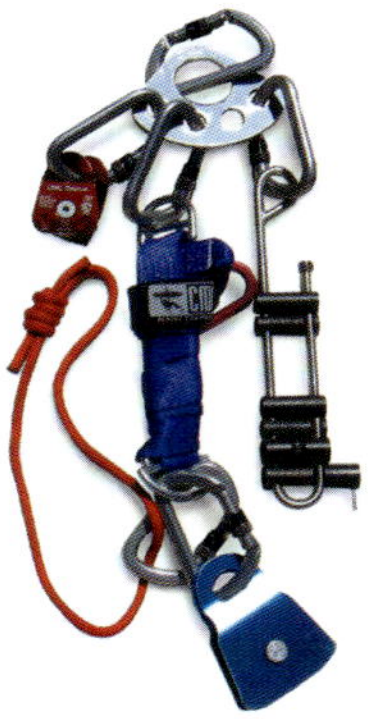

Belay Systems

Whenever rescuers and/or patients are placed on a rope system, it is important to reduce their risk as much as possible. Because there are many ways for a rope system to fail, providing a backup or belay system is the best way to minimize the risk. This consists of a second rope attached to a separate anchor. There are two belay systems in general use: the Tandem Prusik belay and the 540° Rescue Belay.

Set up the belay system close enough to the main lowering system so that there will not be a pendulum if the main system fails, but not so close that it interferes with the operation of the main system. Use a different color rope for the belay to help tell it apart from the main line when communicating rope commands.

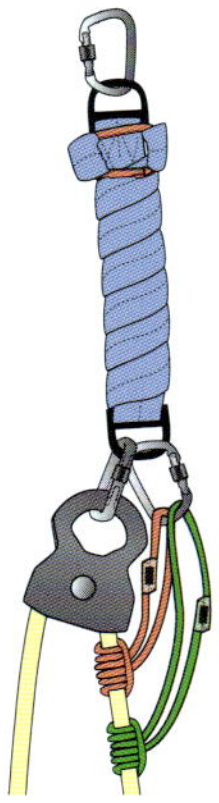

Tandem Prusik Belay Systems

A Tandem Prusik belay system sets up the same for a raising or a lowering belay. Testing has shown that the tandem, triple wrap 8-mm Prusiks provide the best performance for holding large shock loads on 7/16- and 1/2-inch belay lines. A Load Release Strap or mariner's knot is used to attach the Prusiks to the anchor, allowing the load to be released if the Prusiks become locked up during a lowering operation.

The belayer pulls the rope through the Tandem Prusik at the same speed as the raising system. Keeping the slack out of the belay system is still important and a Prusik-minding pulley makes it easier for the belayer to keep up. For a lowering, the belayer tends the Prusik hitches so they do not grab the belay line while the lowering is under control. If the main line fails, the Prusik hitches will be yanked out of the belayer's hands as the belay line catches the load.

Mechanical Belay Systems

The 540° Rescue Belay is a mechanical device that replaces the Tandem Prusik and load release set up. It is a self-locking device that meets demanding drop test criteria: able to stop falling loads quickly while limiting the peak force. It is easy to rig and does not require the ability to tie knots.

A built-in lever is used to release tension on the belay rope, eliminating the need for a release hitch.

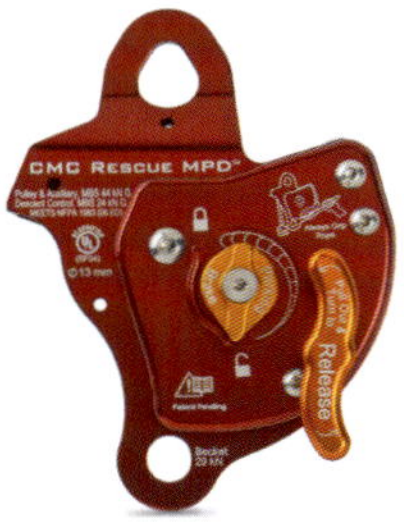

CMC MPD™

Ideal for confined space operations because it converts rapidly to a retrieval line, the patented MPD™ (Multi-Purpose Device) functions as a high-effieciency ratcheting pulley, rescue belay and descent control device (DCD).

Lowering Systems

A lowering system is the most basic system to set up and operate. It is used to lower a load in a safe and controlled manner. A friction device, such as a figure 8 or a brake bar rack, provides braking. The rope is run through the friction device, which is connected to the anchor.

Operation

When both the main and belay systems are rigged, manned and attached to the load, the lowering operation is ready to begin. After the safety check, make sure everyone is ready by using commands. The litter tender, or lowering team leader if there is no tender, starts with the question, "Belay on?" The person manning the belay system replies, "On belay." The tender then asks, "System on?" The person manning the main line replies, "On system." This assures that everyone is paying attention. The tender or team leader then gives the command to begin lowering. Using the command "down rope" instead of "lower" alleviates confusion when adjusting the speed of the operation with the "faster" and "slower" commands. Whatever terminology is used, make sure that all members have a common understanding.

Once the load drops over the edge, communication problems can develop between the tender and the systems. Having an edge-control person to keep visual contact and/or operate a radio can solve these problems and keep the system operators hands free. Developing a backup to the radios is necessary. A simple whistle system, such as the following, works well.

The SUDRH System of Whistle Signals

Stop	1 Whistle
Up	2 Whistles
Down	3 Whistles
Release	4 Whistles
Help	Continuous Blast

The load is controlled by the main system. The belay system follows along, letting the main system handle the load. It is imperative not to allow any slack in the belay system. In the event of a main system failure, slack in the belay system would cause a shock load to the belay system. This shock load could cause the failure of the belay system.

It is important that the people operating both the main and belay systems give their complete attention to the operation. Nobody should be allowed into the area to talk to them while the system is operating. Whether voice or radio communications are being used, if the main system operator or the belayer is not sure of a command, all movement is stopped until the commands are clear.

Raising Systems

A raising system is used to lift a load using a rope. In confined space rescues, powered systems are prohibited. A raising system can be as simple as a single rope attached to the load and pulled by a team of people or a complex combination of pulleys that increases the mechanical advantage (M/A) to a point where one person can raise a load much heavier than their own body weight.

As discussed previously, pulleys perform two distinct functions in mechanical advantage systems. If the pulley is attached to the anchor, it is called a fixed or change of direction pulley. Its job is to change the direction of pull on the rope. If the pulley is attached to the load, it is a movable or mechanical advantage pulley. Its job is to increase the mechanical advantage of the system.

A ratchet is a device that, when attached to an anchor, will hold the rope so that the load will not lower back down when the pulling force is released. This acts as a safety, so the load will not fall back down if the haul team lets go of the rope. This also allows the reset of the mechanical advantage pulleys in order to haul the load further. The number of times that a raising system will need to be reset will depend on the distance the load must be raised and the distance the raising system can be extended. A Gibbs ascender, a Rescucender or a friction knot tied with a Prusik loop can be used as ratchets.

By building systems using different combinations of mechanical advantage pulleys, change of direction pulleys, rope, anchors and ratchets, we can come up with the right tool for the specific job we need to accomplish. Generally, using the lowest mechanical advantage needed to get the job done will result in the quickest rescue.

Simple Pulley Systems

Simple M/A systems can be defined as systems with all the moving pulleys moving at the same speed as the load. A few simple pulley systems are most frequently used. The most common is a 3:1 mechanical advantage, sometimes called a Z-rig because when rigged, the rope looks like a Z. In theory, a 100-pound pull would raise a 300-pound load. Due to friction, the actual mechanical advantage is slightly lower.

Many confined space operations require vertical entry. When we build a mechanical advantage system to move a load in this situation, a simple block-and-tackle system, such as a 2:1 ladder rig or a 4:1 with double pulleys at the top and bottom, can be used.

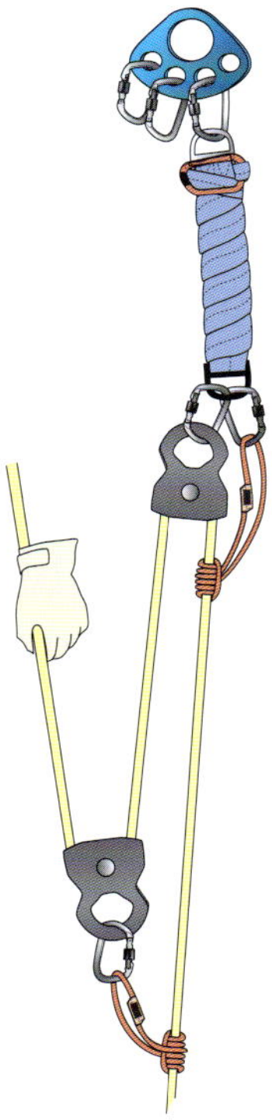

3:1 Simple M/A System

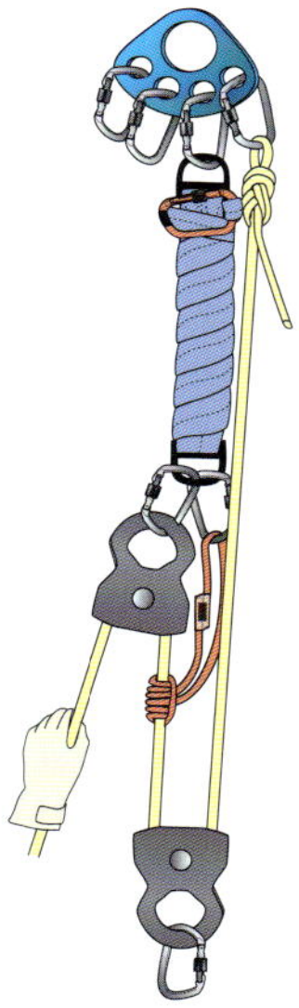

2:1 Simple M/A System With a Change of Direction

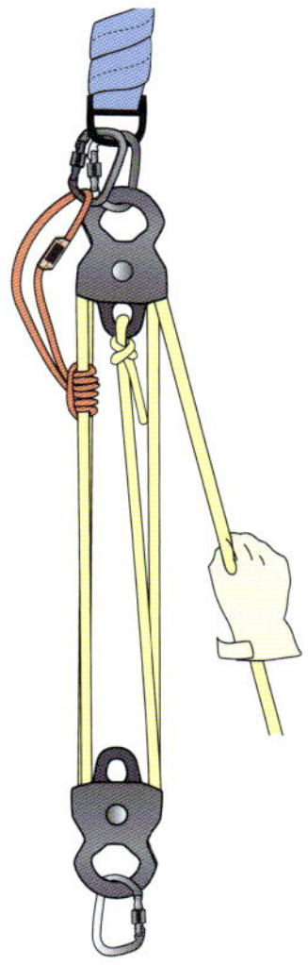

4:1 Simple M/A System

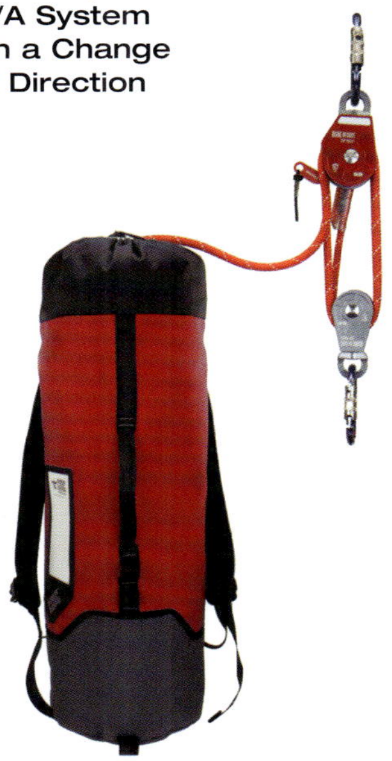

Self-Camming Pulley System

Compound Pulley Systems

Compound pulley systems are created when a simple pulley system is pulling on another simple pulley system. By adding a 2:1 mechanical advantage to a 3:1 mechanical advantage system you compound, or multiply, the mechanical advantage and end up with a 6:1. A 3:1 pulling on another 3:1 gives you a mechanical advantage of 9:1.

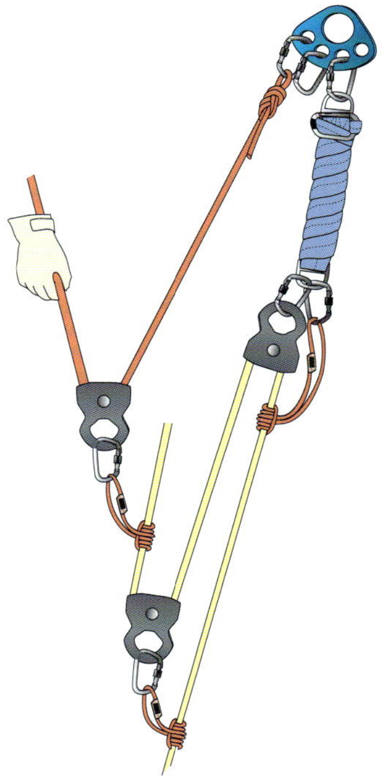

6:1 Compound M/A System

Complex Pulley Systems

A system that is neither simple or compound is a complex pulley system. The 5:1 shown is a complex system.

As the mechanical advantage increases, the amount of rope that must be pulled through the system also proportionally increases. With this in mind, it is best to use as little mechanical advantage as possible to complete the task. A simple 1:1 with a change of direction and a ratchet, with enough manpower to pull the load, will usually get the job done quicker. When a manpower shortage becomes the problem, increase the mechanical advantage.

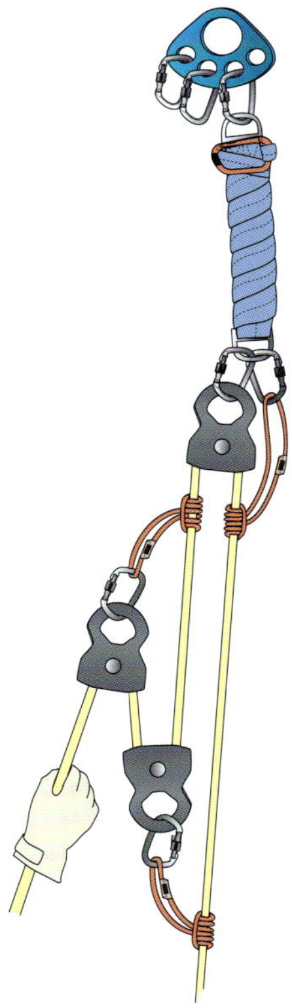

5:1 Complex M/A System

Piggyback Systems

Another method that can be used is to piggyback a mechanical advantage system on a haul line. The load is supported on a separate rope held by a ratchet, and a pre-rigged mechanical advantage system is connected to the haul line with a Prusik hitch when the load needs to be raised. This can be helpful in situations where the load is raised and lowered back and forth a number of times. Piggyback systems can be of any level of mechanical advantage. While not common, piggyback lowering systems can also be rigged. They are usually used for short distances or when knot passing.

The T Method

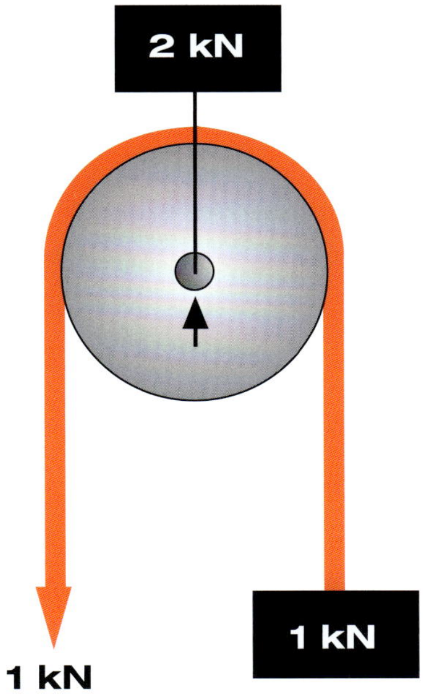

T-Method Pulley

Determining the mechanical advantage of pulley systems can be accomplished in a number of ways. Most are specific to one type of system. The T method works for all systems and provides other important information, such as the maximum force applied to each component of the system. It works by determining the units of tension on each segment of rope in the system. The actual value of one unit of tension varies as the actual input and output forces change. The minimum unit of tension is determined by the load being hauled and by the mechanical advantage of the system. The maximum unit of tension is determined by the maximum input force applied to the system. The input force is determined by the number of people on the haul team.

The basis of the T method goes back to basic pulley theory. If there is 1 kN of force suspended on the end of the rope going through a pulley, there must be 1 kN of force on the other end of the rope to hold it in equilibrium. 1 kN of force on each side puts 2 kN of force on the top of the pulley that is attached to the anchor (or to a Prusik or Gibbs that is connected to a segment of rope) within the pulley system. These units of tension are added to the units that are already on that segment. The sum of the units of tension at the output end is the mechanical advantage of the system.

CALCULATING MECHANICAL ADVANTAGE

1) Only moving pulleys create additional mechanical advantage.
2) Start at the input end of the haul line, the end where the pulleys are working. The unit of tension at this end will be 1.
3) This unit of 1 follows along the rope until the first pulley is reached. If 1 unit enters the pulley, 1 unit must exit. Like the example of the hanging pulley, the addition of both units (in and out) will produce a force of 2 units at the top of the pulley, which, in this example, is the Prusik hitch connection to the load line.
4) Continuing out of the pulley, the rope next enters and leaves a change of direction pulley. Since this is a stationary pulley, nothing is added to the mechanical advantage. Therefore, the unit of 1 comes out of the pulley and moves down to where the Prusik is attached.
5) This is where mechanical advantage is gained. The 1 unit meets the 2 units and both are added together yielding 3 units of tension being applied directly to the load.
6) When comparing the 3 units of tension at the output end to the 1 unit of tension at the input end, a 3:1 mechanical advantage is produced.

THE T METHOD

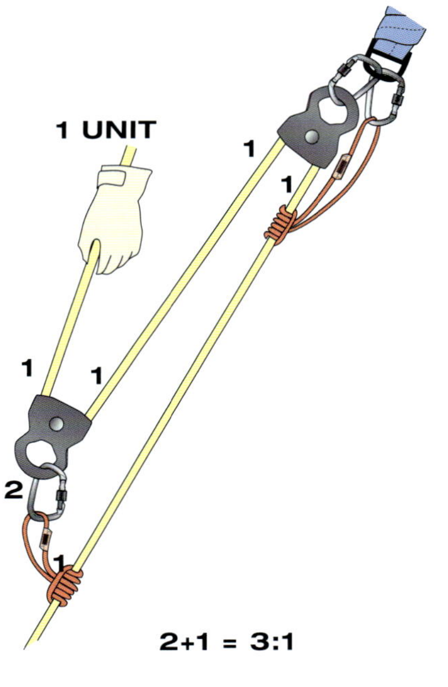

A Simple 3:1 System

**Try this process on the system shown below.
Look at the next page to see
if you worked it out correctly.**

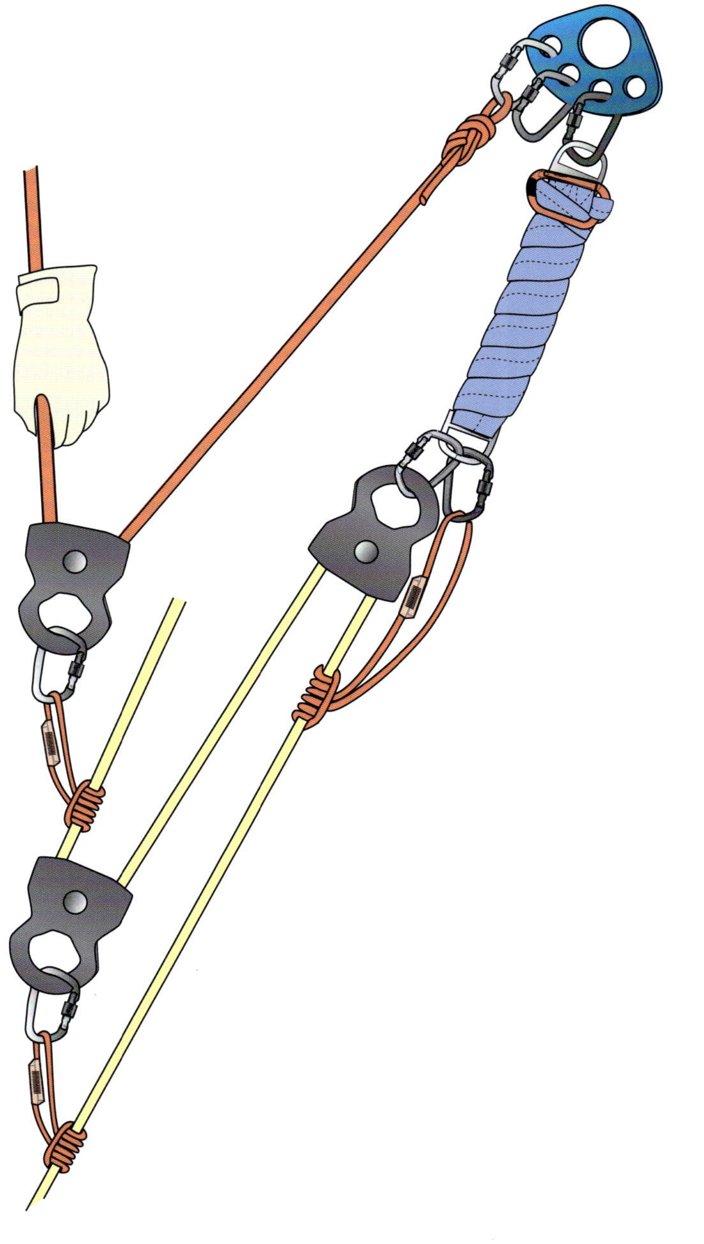

These are the correct calculations for the example on the previous page.

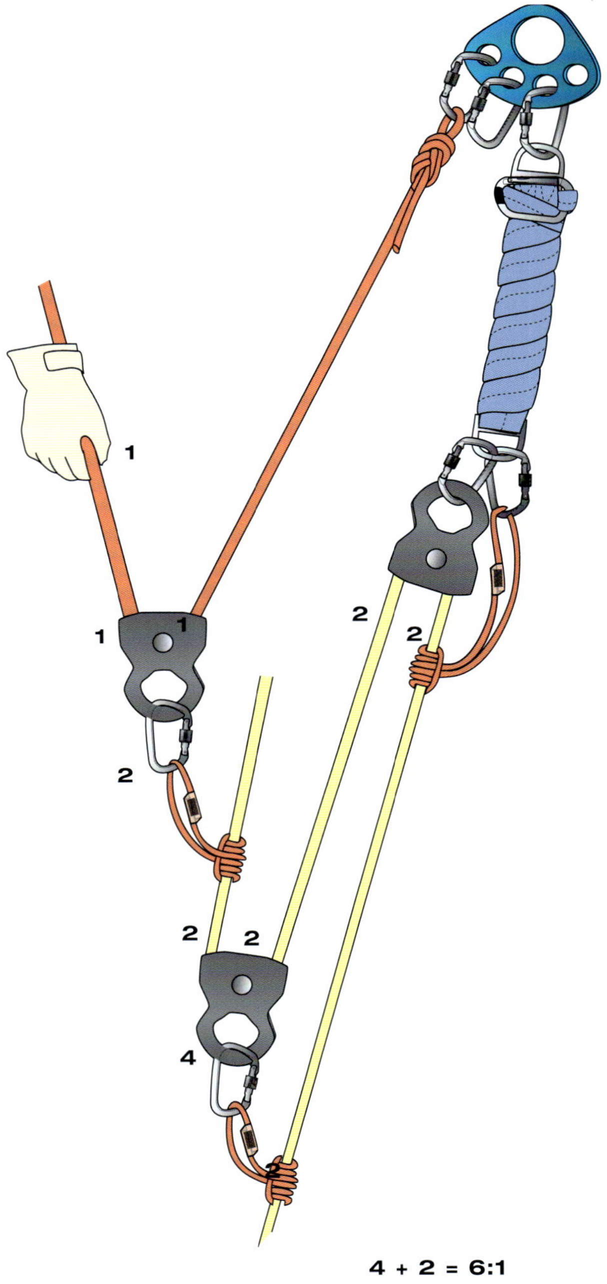

6:1 Compound M/A System

Rope Rescue System Analysis

The analysis of a rope system consists of three parts: the **critical point analysis**, the **whistle test**, and the **white board analysis**. Do a thorough analysis when standardizing team rescue systems. Because the size of the load and the load-bearing capacity of the anchors are often unknown, a quick analysis should also be done on scene. If the rescue requires improvised systems, a full system analysis should be done.

Critical Point Analysis

The critical point analysis looks at each component and asks whether a failure of that component would cause a catastrophic failure of the system. If a critical point is found, it should be eliminated by a redundant component or backed up in some way. Remember to include the belay in the analysis, as it provides the backup for most of the main-line components.

Whistle Test

The whistle test determines what the system will do if all the operators let go at the same time. A swarm of hornets or a bolt of lightning could distract both the haul team and the belayer at the same time. If everyone lets go at once, the load should not fall. The Tandem Prusik belay is an example of a system that passes the whistle test.

Whiteboard Analysis

The whiteboard analysis determines the force on each component based on the way it is used in the system. A rescue system is a chain built from many individual pieces of equipment. The way each component is rigged affects its individual strength, which determines the overall system strength. System safety factor is a function of the weakest link in the system. The whiteboard analysis should evaluate both static loads and dynamic loads.

Static Load Analysis. A white board analysis entails drawing out the system on the board, including any angles, change-of-direction pulleys or other factors that would affect the forces on the system. A simple static system will be used as the first example. Because the forces in kilonewtons provide smaller numbers, the relationships are easier to picture. While the numbers used in the example are approximate, your team's analysis should use the actual strength of each component.

While all systems start with an anchor, the anchor point itself is something that cannot be included in the analysis ahead of time because it is usually different at each rescue location. However, some teams do have anchors that are used repeatedly, such as the tie-off points on the rescue truck. These fixed types of anchors can be included in the analysis. The equipment used to build an anchor system can also be considered beforehand.

A single loop of 1-inch web tied with a water knot around an anchor point has a strength of 21 kN.

The standard aluminum carabiners connecting the rope to the anchor and to the load have a strength of 30 kN. The 1/2-inch (12.5 mm) rope has a strength of 40 kN. Strength loss in the rope will vary with different knots, but in this example we will use 25%, which reduces the rope strength at the knot to 30 kN.

The system load is another factor that cannot be known ahead of time, but a close enough estimate can be made. For this example, we will use 200 kg, a standard load used for rescue tests. A 200-kg load exerts a force 2 kN on the system. Because there are no pulleys or other angles that affect the forces in this simple system, the 2 kN force is exerted on each of the system components.

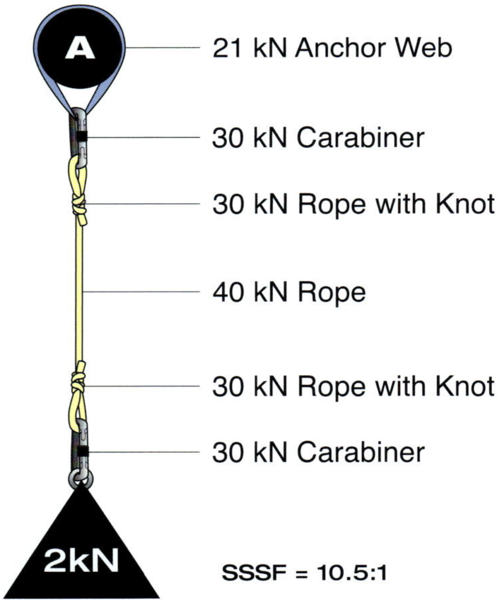

Simple System, First Example

Looking at the system we can determine which component is the weak link, in this case the single webbing loop used to tie the anchor. By dividing the strength of the weak link, 21 kN, by the 2 kN force, we get a Static System Safety Factor (SSSF) of 10.5:1.

To increase the SSSF, make the weak link stronger. By wrapping the webbing around the anchor twice so the load is now supported by a double loop of webbing, the strength of the anchor web increases to 38 kN. This now makes the anchor web the strongest link in the system. The new weak link has a strength of 30 kN and the SSSF increases to 15:1.

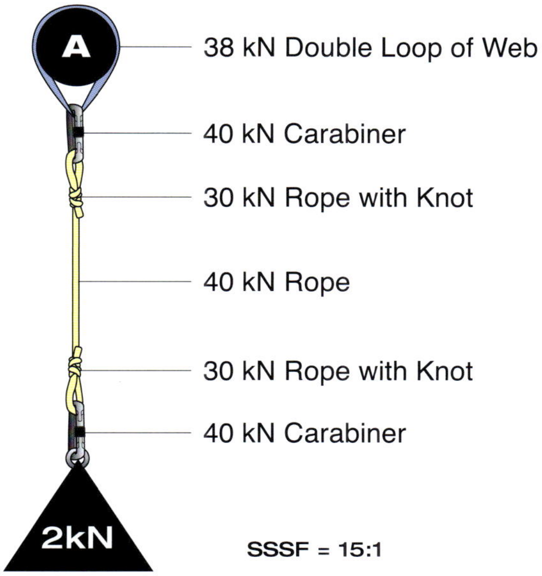

Simple System, Second Example

With this knowledge a rescue team could implement a Standard Operating Procedure (SOP) that requires a doubled loop when tying anchors with 1-inch webbing. If the SOP includes a maximum 90° interior angle for the loop, the anchor web will never be the weak link in the system. Safety in the field has been increased by doing the analysis before putting anyone at risk.

If the system must support the NFPA two-person load of 600 pounds, or 2.67 kN, the SSSF decreases to 11:1. Replacing the 30 kN carabiners with carabiners rated at 40 kN and meeting the NFPA General-Use requirements has not increased the SSSF. The weak link remains at 30 kN, which is the strength of the rope with the knot.

The example on the next page shows a system that might actually be used during a rescue. The change of direction pulley located at the top of the system acts as a force multiplier. If the force on the system is 2 kN, the force on that pulley, carabiner and anchor will be 4kN. (Remember that a change of direction pulley will hold double the load at that point.) If the carabiner holding the pulley is rated at 30 kN than it is the weakest link. Dividing the 4 kN force at that point into the 30kN strength of the carabiner results in a SSSF of 7.5:1. Replacing that carabiner with a NFPA General-Use carabiner at 40 kN makes the 35 kN pulley the weakest link, but the SSSF has increased to 8.75:1. Carrying it one step further and replacing the pulley with a stronger 40 kN model will increase the SSSF to the desired 10:1.

The 30 kN anchor and 35 kN pulley at the second change of direction do not impact the SSSF as much because the rope enters and leaves the pulley at approximately a 90° angle, so the force multiplication is only 1.41 still keeping the system above 10:1. (1.41 x 2kN = 2.82 30/2.82 = 10.63:1 SSSF.)

The resultant forces applied to change of direction pulleys is important to understand when you are using tripods, davits and even ladder trucks as high point anchors and are discussed further in Chapter 10.

Dynamic Load Analysis. Next, consider the effects of dynamic loads. Rescue systems are moving systems and this movement introduces dynamic loads to the system in several ways.

Controlled Dynamic Loading. Controlled dynamic loading occurs during normal operation when the load is lowered or raised. While lowering, gravity is the main influence and a smooth, controlled lower will limit the forces on the system to very close to the weight of the load.

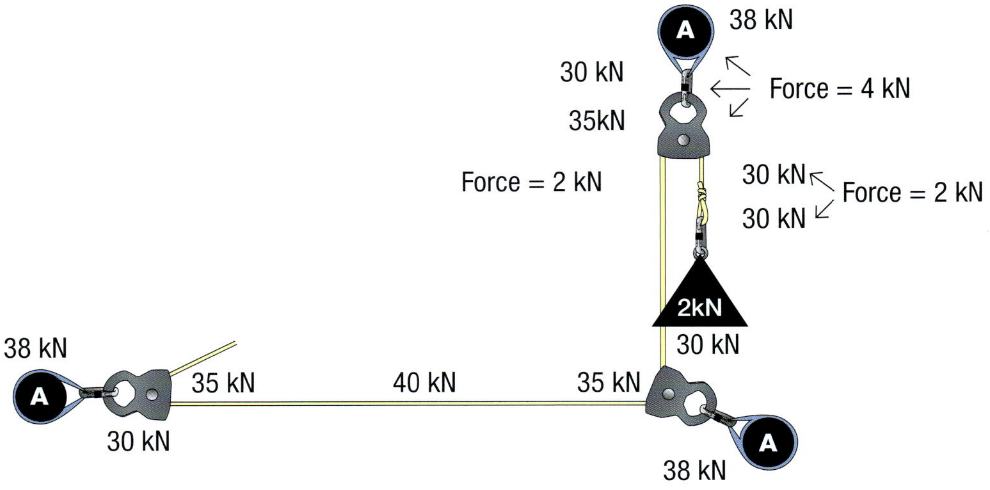

Static System Analysis for a 1:1 Mechanical Advantage System

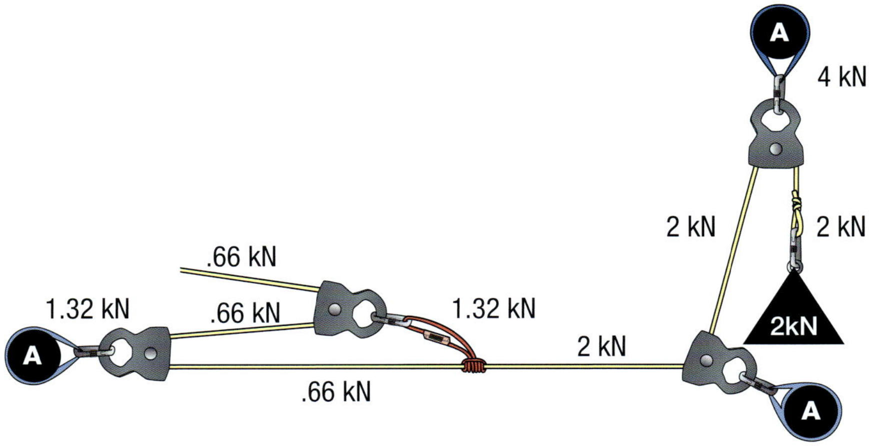

System Forces in a Controlled Dynamic Loading, 3:1 Mechanical Advantage

When raising, the total force on the system is the weight of the load plus any friction caused by the rope dragging over surfaces. The T Method can be used to determine the number of units of tension at each point in the system. The actual force for one unit of tension is calculated by dividing the weight of the load by the total units of tension at the load. By multiplying that value by the number of units of tension at each point in the system, we can determine the force at each at each point.

There is one situation where the weight of the load is not the determining factor of the forces at each point in a raising system. If the load stops moving because it has become stuck or jammed, then forces on the system are determined by the force applied by the haul team. The units of tension can be used to determine the force at each point as a factor of how hard the haul team is pulling.

There are many factors that influence the force of a haul team, such as the strength of the personnel, the gloves worn, and how well their feet can grip the surface. Types of surface such as asphalt, gravel, sand and dirt have a significant effect on a team's ability to pull.

Field tests have provided some average forces that can be applied to a rescue system to determine the potential forces on the system. Using 2.2 kN as one unit of tension for an average three-person haul team should the load jam, use the T Method to determine the potential forces on the system. It should be noted that these forces are not the sustained forces that a haul team applies to a system, but the momentary increase in force as the haul team notices that the difficulty of the pull has increased and then stops to find out why. An untrained haul team may just keep pulling harder without understanding the possibly catastrophic consequences of their actions.

Shock Loads. Uncontrolled dynamic loading or shock loads occur when the load is dropped onto a slack rope. This could happen when a loaded litter is being placed over an edge and then is accidentally dropped before tension is taken up on the rope. The belay line would receive a shock load if a main-line system failure occurred.

The Belay Competence Test Method uses 15 kN as a maximum allowable dynamic load. Applying this 15 kN of force to the system places 30 kN of force on the pulley, carabiner, and anchor at the top change of direction. The resulting DSSF will be 1:1, which is less than a proposed 2:1 DSSF[1]. The actual dynamic forces introduced into the system are impossible to predict and would usually be less. Proper management of rescue systems will limit this loading to manageable levels. Systems can also be engineered to help limit dynamic loading by limiting the amount the load falls before the belay stops it and by also adding shock absorption into the system.

Human Factors

A system analysis is not complete until the human element is considered. Because the human factor cannot be reduced to numbers, it does not become part of the System Safety Factor calculation. Thus, this human factor is often ignored by rescuers interested only in how much load the system can handle.

The human factor must take into account: the level of training and experience of the team, the complexity of the systems and the ability of both the systems and the rescuers to function when conditions become a distraction. Darkness, rain, cold, wind, emotional shock and pressure to hurry (approaching fire, rising water, deteriorating patient) all reduce the rescuer's level of performance. The simpler and more foolproof the systems are, the less chance human error will be the cause of a system failure.

[1] Some authorities have proposed a minimum Dynamic System Safety Factor of 2:1 more as a point to begin the discussion of what factor would be appropriate for rope rescue. Currently an industry consensus does not exist for a minimum Static System Safety Factor, and until the relationships between dynamic and static system performance are better understood, it would be premature to decide on an industry standard minimum Dynamic System Safety Factor.

CHAPTER TEN

HIGH POINT ANCHOR SYSTEMS

SCOPE:

This chapter serves as an introduction to high point anchor systems.

TERMINAL LEARNING OBJECTIVE:

At the conclusion of this chapter the student will be able to identify and describe common high point anchor devices and their related mechanical advantage raising and lowering systems.

ENABLING OBJECTIVE:

- Describe the use of a tripod, including safe loading considerations and common lifting devices or methods
- Describe the various davit arm devices and their advantages and disadvantages
- Identify and explain the correct operation of self-retracting lifeline-type fall protection devices
- Describe the various types of ladder systems that may be used as improvised high points for confined space rescues

There are various types of high point anchors that can be used to assist with entering, exiting and recovering a subject from a vertical opening into a confined space. In some situations a suitable anchor may already exist, such as a structural support for a roof, floor, pipe or machinery. It may be possible to position the lifting point of a crane above the opening. In other cases it may require the use of a manufactured or improvised lifting point, such as a tripod, davit or A-frame. In many cases a temporary or permanent ladder can be used for access, but even then, OSHA regulations require a mechanical method of recovery if the space is greater than five feet. If you are called to perform a rescue and see one of these devices in place, consider it as an option for the rescue. Important points are the integrity of the device, its height and lifting mechanism. Assuming it is not the cause of the rescue, in other words, it isn't collapsed or in pieces on the ground, you could use it, which would save both time and effort.

Tripods

The tripod is the most commonly used high anchor point. It is portable and relatively easy to set up even by one person. Commercial tripods are constructed with telescoping legs that allow it to be collapsed for storage and transport, yet be tall enough to be usable for a rescue. Most tripods are available in two sizes with maximum working heights of seven feet and nine feet. Our experience is that construction and maintenance crews tend to choose the shorter height because it is smaller and lighter (and less expensive.) Remember their use is for ambulatory workers. For rescue, where there is the possibility that you will need to remove a subject on a backboard or in a litter, a taller tripod, nine feet or more, is recommended. If there is an issue with overhead clearance, for instance inside a building with a low roof, the tripod can always be shortened by retracting the legs.

Tripods are free standing and by adjusting the leg extension can be used on inclined surfaces, such as a cone roof tank. Most tripods come with some sort of means to secure the legs from splaying outwards. If a chain, rope or cable is supplied with the tripod, then it must be properly attached to the legs or the tripod may not withstand its rated working load. The legs

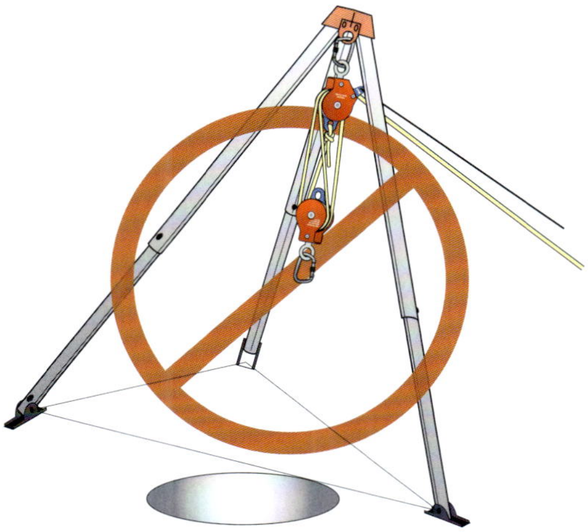

should be angled outward as far as possible for maximum stability. If leg latches are provided, they should be engaged. The tripod should not have to be guyed to remain stable. There are two key points to remember for tripod stability: the lifting means must hang centered within the legs of the tripod and all loads must be kept within the legs. An easy way to remember the later rule is to keep the load within the chain or rope that connects the legs. During a rescue operation it is common for the subject to be hauled from the space and have a well-meaning rescuer pull them away from the opening and tip the tripod over. The feet of the tripod can be staked, tied down to pickets or a similar anchor if the surface allows, but that should not be necessary with proper rigging.

Tripods are strongest when their legs are fully retracted and are required to have a minimum working strength of 310 pounds. That is for fall protection purposes and in most cases the breaking strength is several times that, even with the legs fully extended. That is your indicator that the tripod is meant to be loaded with a single person. It would be a very rare occurrence where both a subject and rescuer would need to be lifted at the same time and that situation should be avoided.

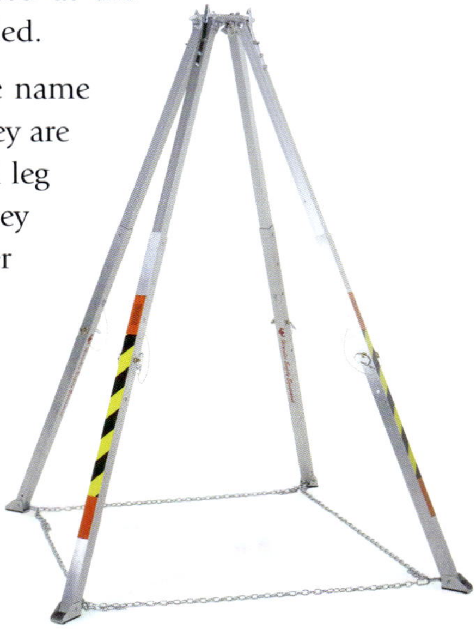

Quad pods are similar to tripods but as the name implies they have four legs instead of three. They are more stable than a tripod due to the additional leg but for the same reason are slightly heavier. They seem to be harder to adjust on anything other than a level surface but the other rules for tripods—securing the legs, centering the load, etc.—still apply. There are not as many options for heights and manufacturers due to their lower popularity.

Arizona Vortex

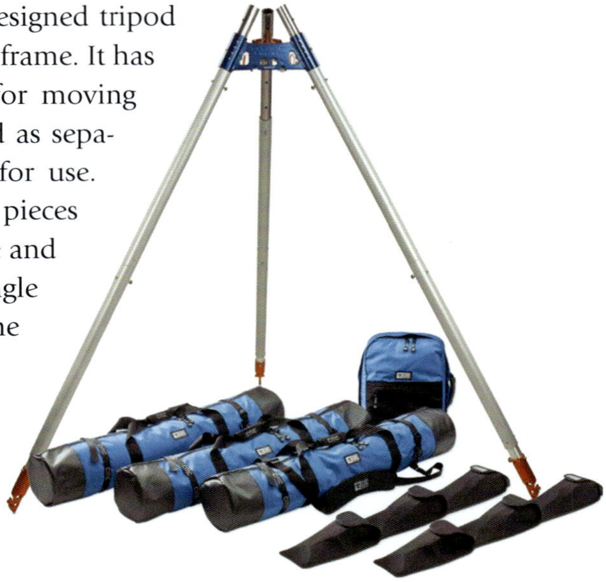

The Arizona Vortex is a specially designed tripod more correctly described as an easel A-frame. It has three support legs that come apart for moving and storage. Easel A-frames are stored as separate pieces and must be assembled for use. That can be an advantage because the pieces are smaller and lighter but set up time and complexity is increased. The fixed angle of the A legs makes the footprint of the easel A larger than a standard tripod, so it might not work in some applications. It can be set up in many different configurations so in some cases, as when a vertical opening is against a wall, it can be used where a tripod wouldn't fit.

Davit Arms

Davit arms are fixed or portable crane-like devices that project over an opening to a confined space. They are often installed in industrial environments, where entries are made on a regular basis or where the height or small work area at the top of a vessel makes the use of a tripod difficult or impossible. Portable davits have an H or V base that has adjusting jack screws to level it. A separate mast and arm assembly is then installed in the base with the tip over the opening. They are usually not as high as tripods, which can make them difficult for rescues, but most have the advantage of swinging back away from over the opening. Another davit arrangement is to have a fixed base socket at each confined space and a single-arm assembly that can be carried between them, depending upon where the work is taking place. This arrangement is common at waste water treatment plants, food processing and paint manufacturing plants or other locations, where there are multiple tanks in close proximity that are entered regularly for cleaning.

Fixed and Mobile Cranes, Ladder Trucks, Etc.

Bridge cranes, mobile construction and maintenance cranes, fire department ladder trucks and other similar devices can be useful as high anchor points for a confined space rescue. They can be positioned over the opening to the space for use as an anchor but must be parked and locked out during the rescue operation. At no time should the winches, hydraulics or other lifting or raising devices be used for the rescue. They are not designed or rated to lift human loads, and the risk that a rescuer or subject could become caught and dismembered before the operator would know and have time to react is too great. Attaching a change of direction pulley for a rope system or hanging a pre-rigged 4:1 and then positioning that over the opening is acceptable. Keep in mind the forces that will be placed on the anchor, including the load multiplier effect of the pulley. Pulling along the beam of the ladder will mitigate that somewhat but do not exceed the manufacturer's recommendations for extension, angles, and tip loads.

Ladder Systems

While tripod or davit arm systems are the preferred high anchor points because of their ease and speed of deployment, ladders can be used in some emergency situations. Ladders can be rigged in two ways. The A configuration uses two ladders and straddles over the opening into the confined space. The ladder gin leans over the space from one side. In either case commercial type 1A or fire service ladders 14 to 24 feet long should be used.

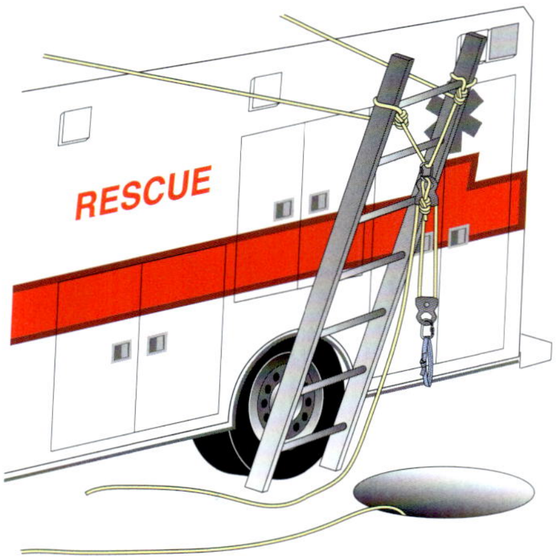

Ladder Gin Against Vehicle

The ladder gin uses a single ladder that leans over the hole and is held at the proper angle by two guy ropes attached at the top of the ladder and extending back and out no more than 45° from the ladder beams. The feet of the ladder must be on stable ground and should be secured to pickets if possible. A rope system, such as the 4:1 pre-rig, is hung from the top of the ladder, which is centered over the opening. As with the other high anchor points, a separate belay system should be constructed and routed at ground level from an anchor to the opening of the space. The system can also be constructed so that the base of the ladder butts into a solid object, such as a curb or even the rear wheels of a fire truck. We have even seen cases where it is built on the tailboard extending out from the rear of the apparatus. If you construct one this way, be sure to pad the feet of the ladder and insure that the guys are tied to structural pieces of the apparatus, not a handrail or mirror.

The ladder A requires two ladders at least 14 feet in length. They can be of unequal lengths and in some cases an extension ladder can be disassembled and the two sections used. If a roof ladder is used, it should be inverted so the hooks are down. The ladder butts are placed together, then spread slightly to allow the beams to be interlocked. The top rungs, or the top rung and adjacent rung if using ladders of unequal length, are lashed together. After the butts are spread to approximately 45°, a guy line is attached to each side of the top of the ladders for stability. The guys are anchored to each side of the ladders and then tensioned. A rope system for raising and lowering and belay is required as with the other high anchors. The butts of the ladders should be either secured together with webbing to prevent them moving apart when the system is loaded, or alternately the butts can be tied to pickets driven into the ground. The potential height possible with a ladder A makes it a good solution for confined space entries and rescues from places such as tanker cars or tank trailers. Those spaces are too tall for a tripod and there is rarely a large enough platform on top to build a tripod.

Ladder A-Frame
with Equal Length Ladders

Ladder A-Frame
with Unequal Length Ladders

Lifting and Retrieval Devices

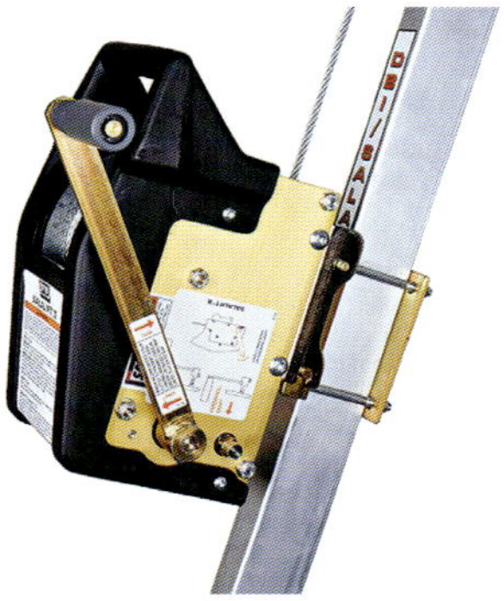

Cable Winches

There are several styles of lifting or retrieval devices that can be used with a tripod, quad pod or davit. The most common, especially in an industrial setting, is a cable winch. They can be equipped with different lengths of cable and various convenience and safety features. The simplest have a fixed handle that is rotated to let out or take in the cable at one speed and level of mechanical advantage. More elaborate models have two speeds, a faster speed with less mechanical advantage for going down and a slower, higher mechanical advantage speed for raising the load. Winches might also include over-speed devices to prevent the load from falling and clutches to stop the handle if the load is too heavy. The later case would occur if the person at the end of the cable became entangled and were not able to tell the person operating the winch to stop.

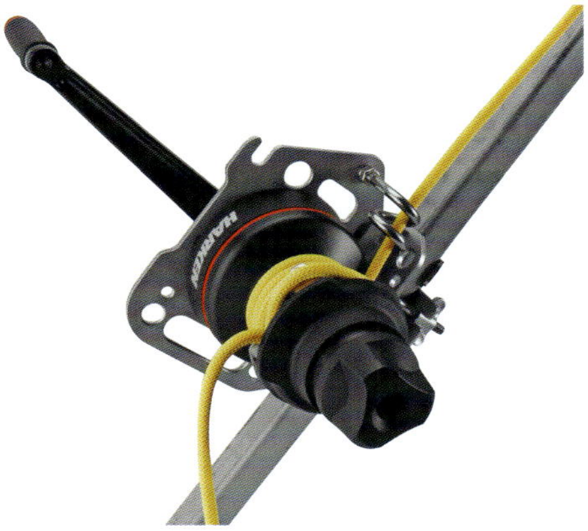

Rope Winches

Two disadvantages to the cable winch are that they can be heavy, especially if loaded with a long length of cable, and in some instances the cable may not be long enough. Rope winches solve those problems yet retain many of the advantages of a winch. They are also designed to fit on the leg of the tripod, and the rope is routed through a change of direction pulley attached to the head of the tripod. The rope is then wrapped several times around the capstan and through a self-tailing device. The most popular have two speeds, which like the cable winch translates to two power levels. For lowering, the rope is released from the self tailer and, using the friction around the capstan for control, is slipped through the operators hands. A single winch can serve several entrants or subjects, which adds safety because they do not need to disconnect from their rope. The operator simply wraps the rope of the person to be lifted around the capstan and cranks them up. By adding a Prusik to the rope that person could even be stopped while suspended in mid lift and another raised up. That would not be possible using a cable winch.

Remember that both the rope and cable winches are meant to be hand-powered devices. In most cases the distance a subject or rescuer needs to be raised from a confined space is not long. By using a powered winch the chance for injury or worse is not worth the saving in time or effort over hand cranking. A winch operator cranking instantly knows when the load increases and is able to stop in time to find out if it is a trapped body part or just the load coming on line.

Pre-Rigged Pulley Systems

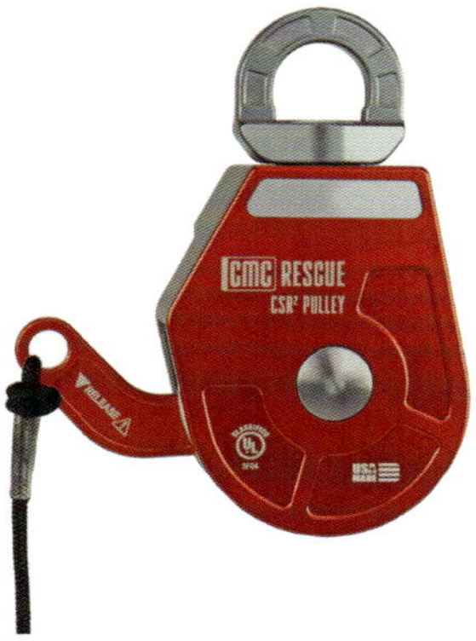

These are rope-based systems that usually use two double-sheave pulleys to achieve a 4:1 mechanical advantage. The attendant or member of the rigging team stands under the top pulley and pulls down to raise the load. The 4:1 mechanical advantage means that to lift a 200-pound load the pull down is approximately 50 pounds, which is easily accomplished by a single puller. For safety some sort of brake system should be installed in the system to prevent the load from falling if the puller lets go. Manufactured systems have the brake built in to the upper pulley, and the systems using larger diameter pulleys and swivels are extremely easy to operate. The upper swivel allows the operator to move anywhere within the tripod area, and the lower swivel means the subject or rescuer can be turned to most easily be pulled out of the space. A suitable length of 1-inch webbing tied to the lower pulley will allow the rescuers in the space to pull the pulley down when making multiple hauls. That is far easier than trying to push the rope through the pulleys. The same technique can be used at the hook end of cable winch if no "headache" weight is attached to help the cable go down.

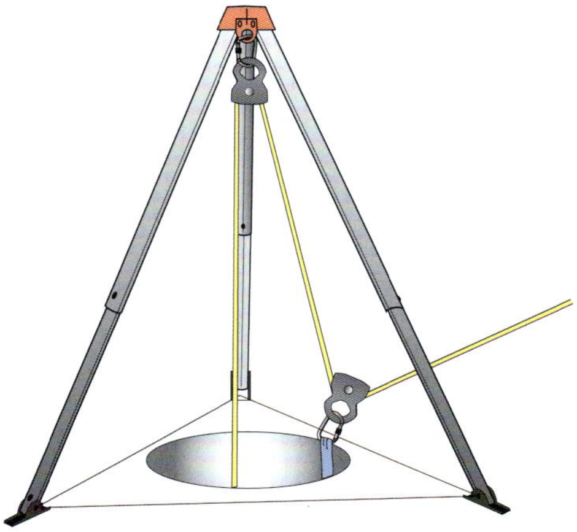

Traditional Rope Systems

Traditional rope-based raising and lowering systems (Chapter 9) can also be used with portable high anchors, but they are less desirable and usually take longer to rig than using a winch or pre-rigged system. A key point to remember is that the pull must be within the footprint of the tripod. If one were to rig the system by hanging a change of direction pulley at the head of the tripod and pulling from a distance away from the space, the tripod will be pulled over. A second change of direction pulley must be installed at ground level within the footprint to prevent that from happening. In most cases that pulley and the rigging needed to attach it gets in the way of the operation or is a trip hazard. There are ways to do it but it takes time and effort. The other reason to avoid these systems is the pulley multiplier effect. The change of direction pulley at the top of the tripod multiplies the force there to up to twice the load. Remember most tripods are designed for a 310- to 350-pound safe working load, and depending upon the weight of your rescuer or subject, you have exceeded that rating.

Fall Protection

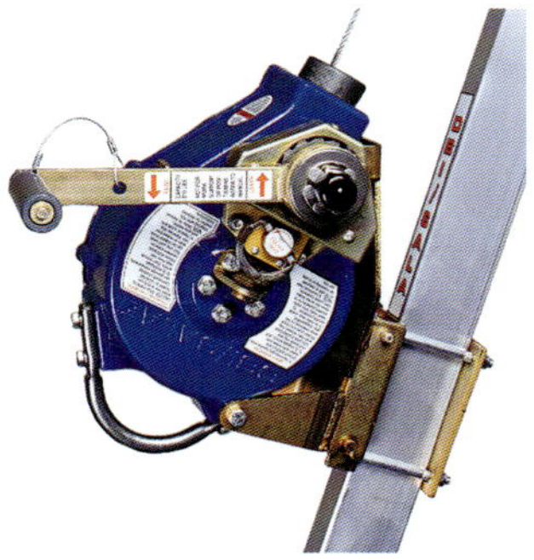

In cases where a portable or permanent ladder is used to access the confined space, a tripod or davit can be used for fall protection. A self-retracting lifeline (SRL) is attached and the cable is routed up through a pulley and attached to the entrant. If they were to fall from the ladder, the lifeline will latch and stop them. The lifeline can be reset by releasing the load from the device (climbing up a few inches) and then they can continue down. If the lifeline has emergency retrieval capabilities, it can also be used for a non-entry rescue. If the entrant becomes incapacitated or non-responsive, the attendant can release the handle and crank the entrant back out the opening. For this feature to be used the attendant must be able to reach the handle, which is impossible if the SRL is hanging from the top of the tripod. It must be at working height with the cable going to the top of the tripod in order to provide enough vertical clearance for the entrant to be fully extricated from the space. Self-retracting lifelines, like some cable winches, contain load indictors that show if the device has been stressed and needs service. Some brands of SRLs need to go for service once the emergency handle has been deployed. That is an expensive process that takes the device out of service for some time. Check the instructions of your SRL before deploying the handle for training purposes.

Belaying

No matter which type of raising and lowering system or high point anchor you are using, a separate rope belay is recommended. The belay will serve as a safety in case of system or anchor failure and for that reason **should be routed at ground level and not through the high anchor point.** If the belay line were routed through a pulley at the top of a tripod and the tripod tipped over, the load would drop twice the height of the tripod before the belay would catch. That fall distance would severely compromise the belay, and in many cases the person would hit bottom before the belay could catch the load. An additional reason for using a Tandem Prusik belay at ground level is that it can easily be converted to a 3:1 (Z-rig) hauling system. Using a belay of this type satisfies two regulations. It provides fall protection and at the same time provides a method for non-entry rescue should an entrant be incapacitated while not on the cable winch or other lifting device.

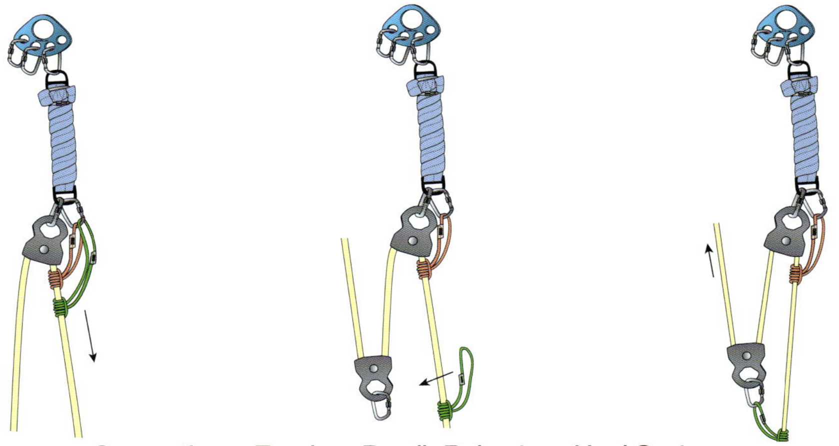

Converting a Tandem Prusik Belay to a Haul System

Resultant Forces on High Anchor Points

The type of device or systems used to raise and lower the load and how it is rigged will determine how much force is applied to the high anchor point. If a winch is attached to the leg of the tripod or the mast of a davit, the force is applied in compression to the upper portion of the tripod leg or davit arm and then vertically down from the point the rope or cable leaves the change of direction pulley. If the load is moved away from vertical, the high anchor point can lose stability.

Tripod with Winch

As stated earlier, rope systems can create different problems where resultant forces are concerned. If the high anchor point is used with a change of direction pulley, the resultant force will bisect the angle formed between the load line and the haul line as they enter and leave the pulley. Since the load line should always be vertical, rigging the haul line to keep the resultant as close to vertical as possible will keep a free standing high anchor point stable.

Tripods are designed to be loaded vertically with the resultant as close to the center of the three legs as possible so that they share the load equally. As the resultant is moved away from center, the load on the leg, or legs, nearest the resultant is increased and the load on the opposite leg is decreased, which reduces the overall strength and stability of the tripod. If the resultant force falls outside the footprint (triangle formed by the three feet or chain around the base,) stability will be lost and the tripod will tip over.

Even with several warnings we commonly see students attempt to pull the subject or rescuer away from the opening after they have been raised out of the space which, if they are allowed to do so, will topple the tripod.

Equally important to the angle of the resultant force is the amount of force different rigging configurations apply to the high anchor point and other change of direction anchors. The actual force is a function of the angle created by the ropes through the pulley. Several examples are shown below.

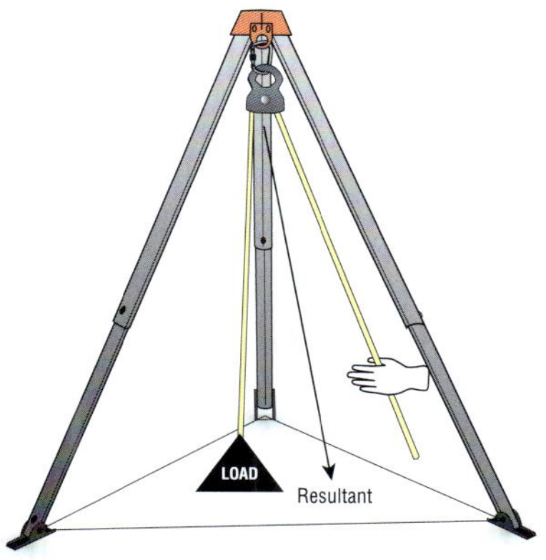

Stable: Tripod with Resultant inside the Footprint

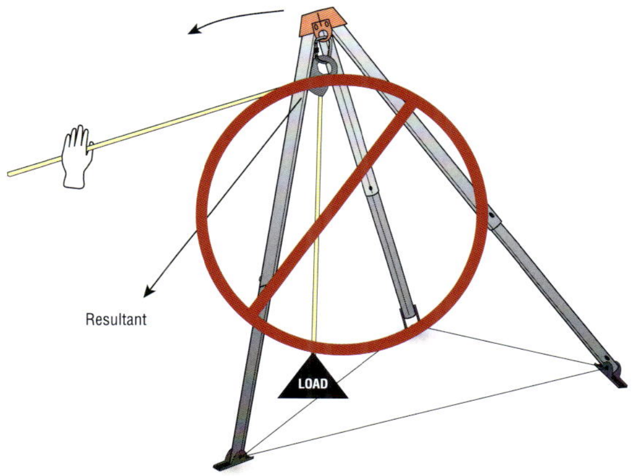

Unstable: Tripod with Resultant Outside the Footprint

Unstable: Tripod with Subject Pulled Away from Opening

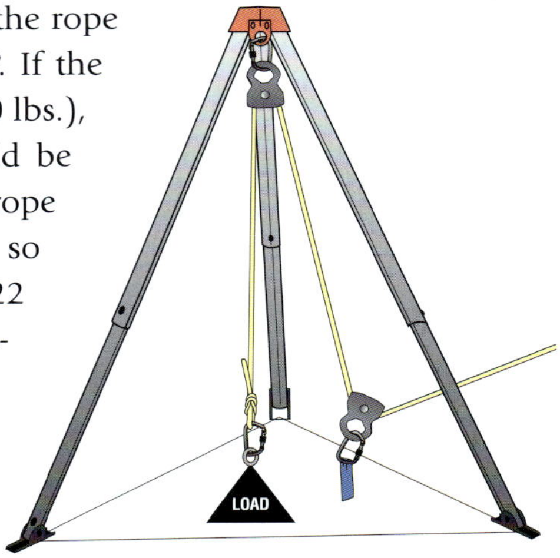

In the example in at right, the angle of the rope through the high directional pulley is 15°. If the load on the end of the rope is 1.33 kN (300 lbs.), the force on the high anchor point would be 2.63 kN (591 lbf.). The angle of the rope through the low directional pulley is 90°, so the force on the lower anchor is 1.88 kN (422 lbf.). And remember, this is a generalization. Due to friction in the pulley, the load will be higher.

Even the commonly used, simple 4:1 pulley system will place a higher force on the high point anchor than the load being lifted. Using the T method to analyze the forces shows that the high point anchor sees 5/4 of the load when hauling. That equates to 1.65kN (375 lbf.) on the anchor when lifting a 1.33kN (300 lbf.) load.

The examples in this chapter show just some of the available tripods, davits, winches, rope and ladder systems available. We have attempted to explain the key points common to most systems, but you should read, understand and follow the manufacturer's instructions for your particular brand of equipment.

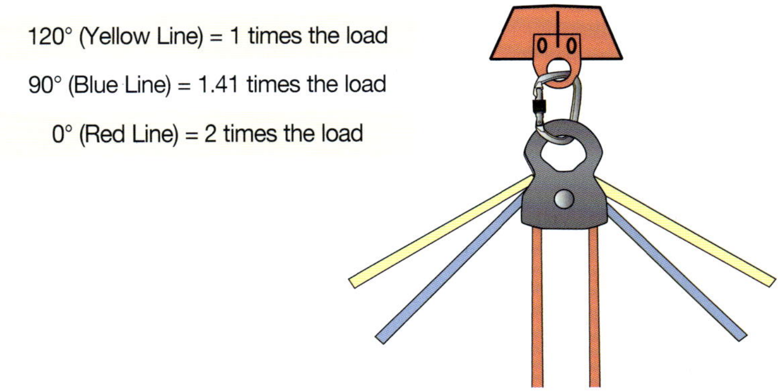

CHAPTER ELEVEN

COMMUNICATIONS

SCOPE:

This chapter serves as an introduction to methods of communication during confined space rescues.

TERMINAL LEARNING OBJECTIVE:

At the conclusion of this chapter the student will be able to describe methods of communicating during a confined space rescue.

ENABLING OBJECTIVE:

- Describe how to effectively communicate using a two-way radio
- Describe how to effectively communicate using rapping and tapping signals
- Describe how to effectively communicate using light signals
- Describe how to effectively communicate using rope signals
- Describe how to effectively communicate using hand signals
- Describe how to effectively communicate using a hard-wired communication system
- Describe how to effectively communicate using using verbal communications

Effective and reliable communications are essential during a confined space entry and rescue operation. Communications are so important that they are required in the regulations. When we discuss communications in this chapter, we are referring to the medium or method from which the message travels from the sender to the receiver. The method used for communication will vary based upon the size and configuration of the confined space, as well as what equipment is available. Regardless of what method of communication is chosen, it must be understood by both the entrant(s) and attendant.

Communications systems used in confined spaces can be typed in different ways but we have divided them into powered and non-powered systems. Powered systems can be divided into those that use wires or those that use radio waves. Non-powered systems include voice, hand, tapping and rapping, rope tugs and light signals. We will describe the various methods by progressing from the simplest to the most complex.

Verbal, or voice, communications are the simplest and easiest. Voice works well in small spaces that do not echo and in cases where breathing apparatus is not being worn. The face piece tends to muffle and distort speech, and while some manufacturers sell amplifiers that can be attached to the mask, echo will still be a problem especially in larger metallic spaces, such as tanks. If voice alone works, the complexity of communications is reduced to zero but you should always have a back up in case it does not.

Hand signals are another simple option for communications. They are not affected by echo but work best where there is clear line of sight between the entrants and attendant. They can be utilized with an intermediate but the additional person in the process will slow things down and can add confusion. Hand signals can really only be used for simple commands, such as those used for raising and lowering systems. It would be just about impossible to give a monitor reading or subject condition using hand signals. We have borrowed these from industry where they have been standardized in lifting operations for many years. Hazardous materials teams have also adopted some of these same signals and added one of their own. The signal indicating the entrant is low on air and needs immediate attention is one or both hands in front of the throat in a simulated choking motion.

Commonly Used Hand Signals

Raise Load

Lower Load

Stop

Change Over

Low on Air

Help

Rope signals are another simple form of communication that utilize material already in place during the rescue operation. It can either be the rescue line or a smaller tag line that is in place only for communication. The acronym **OATH** has been widely adopted as a guide for simple messages.

O = OK	1 tug indicates all is OK
A = ADVANCE	2 tugs tells the other person to advance the retrieval line
T = TAKE UP	3 tugs tell the other person that the line needs to be taken up
H = HELP	4 tugs tells the other person that they need help or to evacuate the space

Light Signals use a common hand-held flashlight as the signaling device. A rescuer would usually have both a headlamp and a hand lamp with them but the hand lamp is probably easier to use for signaling. As with the rope tugs the acronym **OATH** is used as the foundation for transmitting and receiving light signals.

While the equipment is simple, light signals also require line of sight between the entrants and attendant or they will not work. The other difficulty is that it is hard to determine when the entrant is signaling as opposed to the natural movement of their head and headlamp while they are moving in the space.

O = OK	1 flash indicates all is OK
A = ADVANCE	2 flashes tells the other person to advance the retrieval line
T = TAKE UP	3 flashes tell the other person that the line needs to be taken up
H = HELP	4 flashes tells the other person that they need help or to evacuate the space

Rapping and Tapping codes are another option for confined space communication. They do not require line of sight and depending upon the type of space can travel long distances. They might trace their origin to mining where tapping, or more correctly, banging on timbers was used by trapped miners to tell their rescuers they were alive and give them an approximate location. Our method uses a carabiner or other metallic object to tap on the side of the tank. A predetermined set of codes must be understood before entry but, like the others we described, **OATH** is an easy acronym to use.

O = OK	1 rap on the side of the vessel indicates all is OK
A = ADVANCE	2 raps tells the other person to advance the retrieval line
T = TAKE UP	3 raps tell the other person that the line needs to be taken up
H = HELP	4 raps tells the other person that they need help or to evacuate the space

There are several disadvantages to using the rapping/tapping method of communication. Some confined spaces are constructed of, or lined with material, that does not transmit or dampens sound. The converse of this is that some materials transmit sound over long distances. That means you may be hearing noise from a process that is not part of the rescue operation. Tanks and other spaces with large volumes tend to echo, which makes counting the raps difficult, and the act of crawling through a space with a harness full of carabiners can simulate the rapping and tapping signals. A final concern with this method of communication is the atmosphere in the space. If there is the possibility of a spark from a rapping or tapping tool igniting the atmosphere, it should not be used.

Portable Radios

Handheld portable radios are a form of communication that rescuers should already be familiar with. There are no acronyms to remember, they are easy to operate and they are probably part of the rescue team's equipment cache. The frequency on which the radios operate and the configuration of the confined space both play a key part in how well the radios will operate for communications between the attendant and entrants. The higher the frequency, the more the radio depends upon line of sight for proper operation. Radios in the UHF bands will often fail to receive the signal if it needs to go around as few as one or two corners. Lower frequency VHF radios perform a little better but can still be limited by the configuration of the space.

Another key point to remember is that the radios should be switched from conventional or repeat mode to direct or talk around. In the former the radios transmit to a repeater, which in turn rebroadcasts the signal to the receiving radios on a different frequency. In most cases any radios inside of the space would be effectively shielded and therefore their transmissions would not be able to reach the repeater for rebroadcast.

Some other considerations with using radios for confined space communications are the tendency for face pieces to garble voice when wearing breathing apparatus. That difficulty can be eliminated by using face pieces with microphones that plug into the radio or with throat microphones. Another problem is that the radios work with a push-to-talk system to turn on the transmitter. That means that you must have a hand free to access the radio to push the switch to talk. Sometimes that can be difficult when negotiating a small confined space or when trying to talk while doing something with your hands.

A final caution when using radios is that they be properly rated for the atmosphere in which they are being used. Some radios are tested for use in explosive atmospheres, others aren't. That includes any accessories, such as remote microphones or headsets, that are taken into the confined space.

Wired Communications Systems are generally considered to be the best way to communicate during a confined space rescue. They are self contained so they do not require a repeater and do not interfere with other radio traffic. Some systems operate by using a device similar to a handheld radio that slides over a wire that serves as the antenna. Several units can be attached and they have accessories that allow them to work with breathing apparatus. The disadvantage is that they are a little bulky and require an available hand for push-to-talk operation. The most popular system uses a control box with the attendant at the entrance to the space and is configured so that all the entrants are connected. They can easily talk among themselves in full duplex mode, which means they operate like your telephone—both entrant and attendant can speak at the same time without using a push-to-talk switch. Small earpieces of various styles fit under the entrant's helmet and throat mics help insure good sound quality without specialized face pieces.

Sound-powered systems are also available. As the name implies these do not require batteries but do require the wire, and as users are added to the system the volume is reduced. The other inconvenience with the sound-powered systems is that the entrant headsets are bulky and uncomfortable to wear with breathing apparatus. The small headsets and throat microphones available with the powered systems are not available, so speech is sometimes difficult to understand because it is heard through the diaphragm of the mask.

Wired systems are expensive and are a specialized piece of gear but for non line of sight, long distance, complete communications, they are the recommended method. The most common objection to wired systems is the wire. The entrants must drag it with them and there is the potential for cuts, disconnections and entanglements. While all are true, we have not seen most of those problems in our training classes. In most cases the entrants will have either an air line or a rope tag line with them, so the communications line does not really become an additional hazard or inconvenience. We have encased our communications cables in plastic sheathing along with the air lines, so there is only a single hose to manage. Some manufacturers sell their communications cables braided into G-rated kernmantle rope. That combination also reduces the number of lines entering the space. We prefer the communications/air line combination because it can serve as a tag line and the regulations do permit removing a retrieval line if it hinders movement or will not aid in rescue. However, this would be impossible if the rope contained your communications line. As with radios, it is important that any powered communications system be rated for the atmosphere in which it is to be used.

We suggest two methods for cable management. As mentioned earlier, we bundle our communications and air lines together in a plastic sleeve. Prior to that we used 2-inch webbing but it is bulky, hard to clean and snags easily. Even before that we simply used colored electrical tape. If we put the tape at a standard interval, we could estimate distance but it seemed that the tape would wear and fall off quickly. Outside the space you can either stage the lines in a large figure 8 pattern or we pack ours in plastic shipping containers using the over/under coil. The figure 8 and over/under both keep the twists out of the cable, which either snags or kinks, possibly shutting off the air flow.

Figure 8 Coiled Hose

Over/Under Coiled Hose

There may be times when wired or radio communications will not work or is not available, so it is always best to have a backup method in place and fully understood by the entrants and attendant.

Components of a Typical Confined Space Communications System

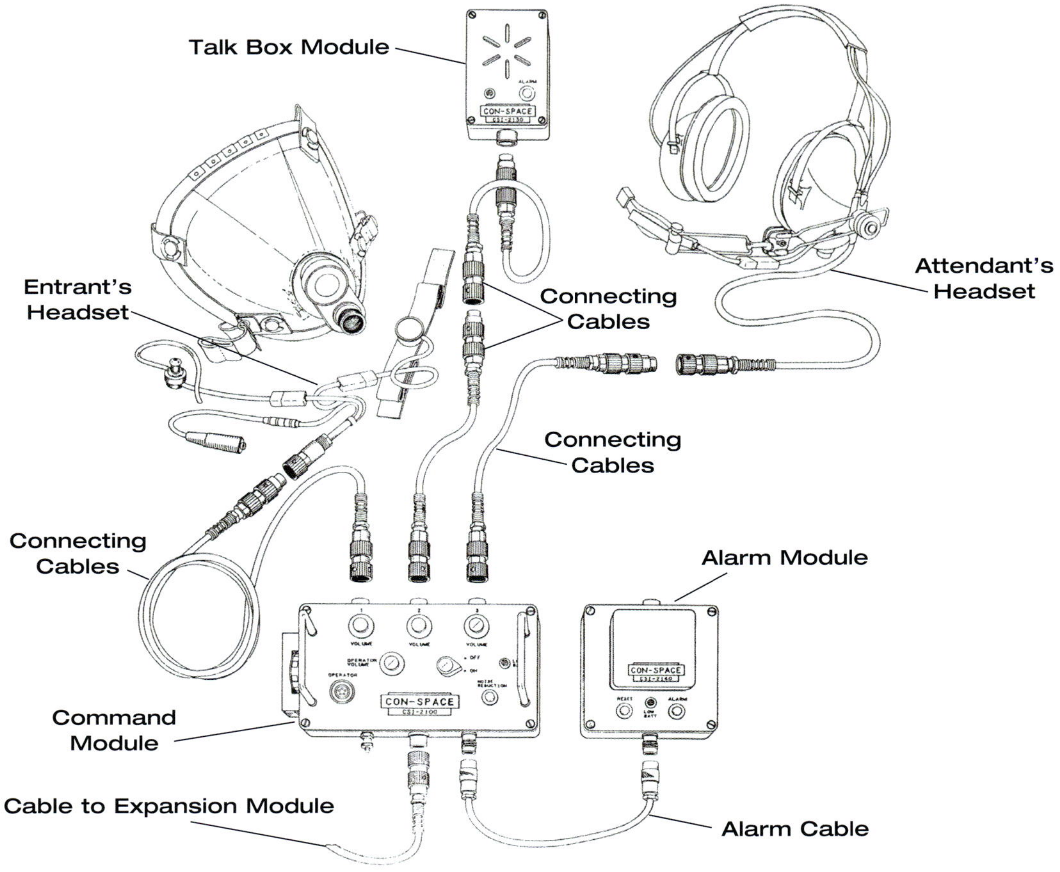

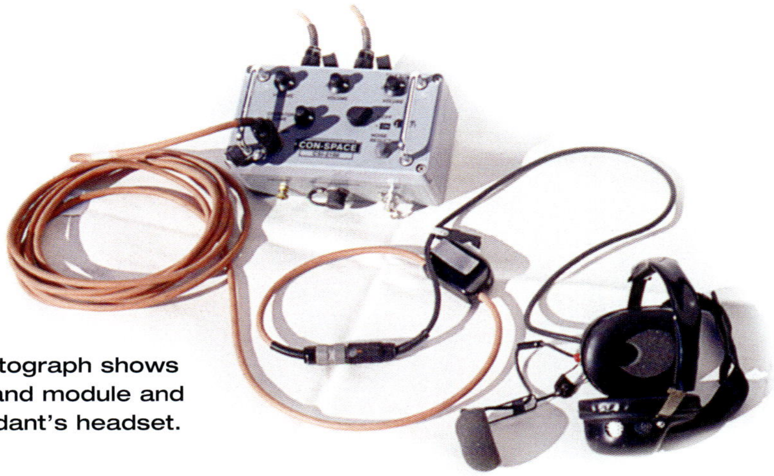

This photograph shows a command module and an attendant's headset.

CHAPTER TWELVE

PERMITTING CONFINED SPACES

SCOPE:

This chapter serves as an introduction to the confined space permitting process.

TERMINAL LEARNING OBJECTIVE:

At the conclusion of this chapter the student will be able to complete a permit for a confined space entry.

ENABLING OBJECTIVE:

- List the reasons for permitting a confined space
- List the 15 requirements of a confined space entry permit
- Describe the local agency policy for when a permit is issued and its duration

Just as the name implies, OSHA regulations require that a permit be completed before entering a permit-required confined space. That regulation holds true for rescuers, as well as the original entrants into the space. The reasoning behind this requirement is to insure that all hazards are identified, controlled, special access equipment is identified and that training and potential rescue needs are identified and in place. Remember, OSHA wrote the regulations in response to incidents and their mandate is to prevent worker injuries and fatalities. The permit itself serves as a checklist to be sure that everything is taken care of before anyone enters the space once again to try and prevent accidents. The checklist or tactical worksheet concept is especially valid for public safety agencies that don't make entries or perform rescues in confined spaces on a regular basis. While permits are only required for permit-required confined spaces, we encourage their use whenever the rescue team makes an entry. The safest policy is to view all spaces as permit required until all hazards are thoroughly evaluated and eliminated. We, as rescuers, are called because something went wrong somewhere in the process, and we should err on the side of caution until we have all of the information we need to safely enter the space.

The permit has several purposes beyond the overall goal of injury prevention.

- It serves as documentation that all of the steps necessary for safe entry into the space were competed
- It serves as a safety checklist
- It serves as a tactical log of what happened and when
- It provides an incident action plan, size up and safety-related information

Remember, the permit and the other documentation used during a confined space rescue become legal documents. Permits must be kept on file for one year.

Sample Confined Space Rescue Permit (page 1 of 4)

This permit shall be completed in its entirety, remaining at the rescue site for the duration of the rescue operation and kept on file for one year following the event. Shading denotes an operational priority or mandatory component.

INCIDENT NUMBER	INCIDENT NAME	DATE/TIME
INCIDENT LOCATION		
RESCUE START DATE AND TIME	RESCUE END DATE AND TIME	
DESCRIPTION/USE OF CONFINED SPACE	FACILITY CONTACT	
SPECIAL POTENTIAL HAZARDS		

ICS Assignments

RESCUE GROUP SUPERVISOR	ATTENDANT
AUTHORIZED ENTRANT #1	BACKUP ENTRANT #1
AUTHORIZED ENTRANT #2	BACKUP ENTRANT #2
AIR MONITORING	SAFETY OFFICER
AIR SUPPLY	RIGGERS

Pre-Entry Checklist

- ☐ Operations Perimeter Setup
- ☐ Atmospheric Monitoring
- ☐ Ventilation
- ☐ Eliminate Ignition Sources
- ☐ Confirm Lockout/Tagout
- ☐ Provide Lighting
- ☐ Respiratory Protection
- ☐ Protective Clothing
- ☐ Communications
- ☐ Pre-Entry Briefing

Communication Plan

- ☐ Visual/Hand Signals
- ☐ Voice
- ☐ Radio (Intrinsically Safe)
- ☐ Rope Signals (OATH)
- ☐ Hardwire

* Identify Backup Communication Plan

Ventilation Plan

- ☐ Natural
- ☐ Forced Exhaust
- ☐ Forced Supply
- ☐ Microatmosphere
- ☐ Other:

A full-size version of this form can be found in the appendix.

Sample Confined Space Rescue Permit (page 2 of 4)

ENTRANT	ENTRY TIME	SCBA PRESSURE	SAR PRESSURE	EXIT TIME

GAS	PHYSICAL CHARACTERISTICS	FLAMMABILITY LEL	TOXICITY	SYMPTOMS
Carbon Monoxide CO	Colorless Odorless	12.5% 125,000 ppm	IDLH 1,200 ppm	Headache, Nausea, Dizziness, Tachypnea
Carbon Dioxide CO_2	Colorless Odorless	Non-Flammable	IDLH 40,000 ppm	Headache, Dizziness, Restlessness, Sweat, Dyspnea
Methane CH_4	Colorless Odorless	5%	Simple Asphixiant	Blurred vision, Cough, Tachycardia, Tachypnea
Hydrogen Sulfide H_2S	Colorless Rotten-Egg Odor	4% 40,000 ppm	IDLH 100 ppm	Eye Irritation, Respiratory Irritation, Headache
Sulfur Dioxide SO_2	Colorless Suffocating Odor	Non-Flammable	IDLH 100 ppm	Eye, Nose, Throat Irritation, Coughing, Skin Burns
Nitrogen Dioxide NO_2	Yellow/Brown Pungent Odor	Non-Flammable	IDLH 20 ppm	Cough, Frothy Sputum, Eye Irritation

Notes:

A full-size version of this form can be found in the appendix.

Sample Confined Space Rescue Permit (page 3 of 4)

Prior to entry a pre-entry briefing shall be performed with all key personnel, this includes, but is not limited to:

☐ ENTRANT	☐ ATTENDANT
☐ BACKUP ENTRANT	☐ RESCUE GROUP SUPERVISOR

The following information will be reviewed prior to entry:

- ☐ The hazards that may be encountered specific to this entry. (Atmospheric, Engulfment, Mechanical, Physical, Corrosive, Biological)
- ☐ The primary and backup communications plan.
- ☐ A confirmation that the entrant has all equipment needed to perform a successful entry and is trained on all of the equipment. (PPE, Respiratory Equipment, Communication, Rigging, Patient Packaging)
- ☐ A review of any potential self rescue plans if possible.

ENTRY AUTHORIZED

Rescue Group Supervisor:

Signature: Date/Time:

Notes:

ENTRY CANCELLED

Rescue Group Supervisor:

Signature: Date/Time:

A full-size version of this form can be found in the appendix.

Sample Confined Space Rescue Permit (page 4 of 4)

Atmospheric Monitoring Results

DATE	TIME	LEVEL	% OXYGEN	% LEL	(PPM) H_2S	(PPM) CO	INITIAL

Name (print):	Signature:	Date/Time:

A full-size version of this form can be found in the appendix.

A confined space entry permit must contain the following information:

1. The permit space to be entered
2. The purpose of the entry
3. Date and duration of the permit
4. The names of the authorized entrants and a system to track them for the duration of the permit
5. The name of the attendant(s)
6. The name of the entry supervisor(s), with the signature or initials of the entry supervisor who originally authorized entry
7. List of hazards of the space
8. List of hazard controls necessary before entry
9. List of acceptable entry conditions
10. Recorded test results with the time tested and the names or initials of the testers
11. Rescue and emergency services that are provided on site
12. Communication procedures for use by attendants and entrants
13. All equipment needed for the entry
14. Any other information specific to the entry
15. Any other permits necessary for the work authorized (such as hot work permits)

The duration of the permit will be based upon your agency procedures. It may be until the rescue is completed, it may be when there is a crew change for a long duration operation or it may be based upon a fixed amount of time.

The rescue permit, described in Chapter 8 and shown here, contains similar information but is altered slightly considering the circumstances. It has a box to indicate if a confined space entry permit is available but that is often not the case. The theory being that by following the permit process the need for rescue would be almost eliminated. The rescue permit follows the phases of rescue and is supplemented by the IC chart, tactical and air supply log (if used) for a more complete documentation package.

The concept of completing a permit before beginning a rescue is a little difficult for many rescuers to grasp. We keep going back to the origins of the regulations: to keep us alive. Upon review of the statistics, the number of rescuers (or would-be rescuers) killed continues to exceed the number of original subject fatalities. It only takes a few minutes to complete the paperwork and by following the process you may be saving your life or that of your fellow rescuers.

APPENDIX

SAMPLE CONFINED SPACE RESCUE FORMS

Confined Space Survey Form

Confined Space Entry Permit

Confined Space Rescue Permit

Confined Space Tactical Log

Confined Space Air Supply Log

Confined Space Rescue Incident Command System

Confined Space Rescue Permit

Rescue Team Evaluation Criteria

Confined Space Survey Form

CONFINED SPACE NUMBER	PERMIT REQUIRED ☐ YES ☐ NO	DATE OF SURVEY

☐ Alternate Entry Procedures Possible
☐ Atmospheric Hazards Only ☐ Permit-Entry Required for Testing ☐ Eliminated by Ventilation

☐ Reclassification as Non-Permit Space Possible
☐ No Actual or Potential Atmospheric Hazards ☐ All Other Hazards Eliminated
☐ Permit-Entry Required for Hazard Elimination

LOCATION OF SPACE

DESCRIPTION OF SPACE

CONFIGURATION OF SPACE

POSSIBLE ATMOSPHERIC HAZARDS
☐ Oxygen < 19.5% ☐ Oxygen > 23.5% ☐ Flammable Atmosphere ☐ Toxic Atmosphere

POSSIBLE GASES PRESENT

PHYSICAL HAZARDS
☐ Mechanical ☐ Electrical ☐ Pneumatic ☐ Hydraulic ☐ Pipes or Ducts
☐ Engulfment ☐ Hazardous Substance ☐ MSDS Required TEMPERATURE: ☐ High ☐ Low

UNUSUAL HAZARDS

REASONS FOR ENTERING SPACE

WHO USUALLY ENTERS SPACE

NUMBER OF ENTRY POINTS	FREQUENCY OF ENTRY

EXTERNAL CONNECTIONS TO SPACE

SURVEY COMPLETED BY

Print Name Signature

Provided by CMC Rescue, Inc. Copies of this form can be downloaded at cmcrescue.com/forms.

Confined Space Entry Permit

CONFINED SPACE NUMBER	DATE OF ENTRY	DURATION OF ENTRY

PURPOSE OF ENTRY

ENTRANTS

Print Name ____ Signature ____ IN ____ OUT ____

Print Name ____ Signature ____ IN ____ OUT ____

ATTENDANT(S)

Print Name ____ Signature ____ IN ____ OUT ____

ENTRY SUPERVISOR(S)

Print Name ____ Signature ____ IN ____ OUT ____

ATMOSPHERIC MONITORING

ACCEPTABLE ENTRY CONDITIONS	TIME								
> 19.5% < 23.5%	OXYGEN								
< 10% LEL	FLAMMABLE								
PEL _______ ppm PEL _______ ppm									
CONTINUOUS MONITORING	TESTER'S INITIALS								

HAZARDS

☐ Mechanical ☐ Electrical ☐ Pneumatic ☐ Hydraulic ☐ Pipes or Ducts ☐ Engulfment

TEMPERATURE: ☐ High ☐ Low ☐ Hazardous Substance ____________

Other Permits Needed: ☐ Hot Work ☐ Safe Work

HAZARD CONTROL

☐ Ventilation Fans: Duct _____ feet ☐ Saddle Vent with Bend MECHANICAL: ☐ Block Linkage ☐ Disconnect

ELECTRICAL: ☐ Lockout ☐ Tagout PIPES/DUCTS: ☐ Blind ☐ Disconnect

HYDRAULIC: ☐ Disconnect ☐ Lock Pump and Bleed PNEUMATIC: ☐ Disconnect ☐ Lock Compressor and Bleed

☐ ____________

PROTECTIVE EQUIPMENT

LIGHTING (EXPLOSION PROOF): ☐ Cord Light ☐ Caplamp ☐ Generator ☐ Power Cords: _____ feet

☐ Fire Extinguisher ☐ Static Protection ☐ Warning Signs ☐ Cones ☐ Barricades/Tape

COMMUNICATIONS: ☐ Hardline ☐ Radios ☐ Hand Signs ☐ Rope Signals

☐ Tripod ☐ Davit ☐ Winch ☐ Retractable Lifeline with Winch ☐ Ropes ☐ Hardware ☐ Ladder

PERSONAL PROTECTIVE EQUIPMENT

☐ Body Harness ☐ PAL ☐ Safety Glasses ☐ Goggles ☐ Face Shield ☐ Hard Hat ☐ Ear Muffs/Plugs

☐ Coveralls ☐ Leather Gloves ☐ Splash Suit ☐ Chemical Gloves ☐ Overboots

RESPIRATORY PROTECTION

☐ SAR ☐ SCBA ☐ Cascade ☐ Airline: _____ feet ☐ APR ☐ PAPR

CARTRIDGE: ☐ Organic Vapor ☐ Acid Gas ☐ Organic Vapor/Acid Gas ☐ HEPA ☐ Ammonia ☐ Dust/Mist

RESUCE AND EMERGENCY SERVICES

Rescue Team: ____ Phone No./Radio Call Sign: ____

Location of Phone: ____ Radio Frequency/Channel: ____

RESCUE EQUIPMENT: ☐ SKED ☐ Backboard ☐ Half Back

Were any unexpected hazards encountered during the entry? ☐ No ☐ Yes (explain below)

TIME PERMIT CANCELLED ____ PRINT NAME ____ SIGNATURE ____

Confined Space Rescue Permit

ASSESSMENT

DATE	LOCATION	
TIME		
RESPONSIBLE PARTY/CONTACT PERSON		
NUMBER OF VICTIMS	TIME LAST SEEN	CONDITION

☐ ENTRY PERMIT AVAILABLE

DESCRIPTION OF SPACE	ACCESS
CONTENTS OF SPACE	☐ MSDS AVAILABLE

HAZARDS IN SPACE

☐ Mechanical ☐ Electrical ☐ Pneumatic ☐ Hydraulic ☐ Other

ATMOSPHERIC: % Oxygen % LEL TOXICITY: % ppm of

Other Toxic Substances: % ppm of % ppm of

TIME TESTED	PERSON TESTING	METER CAL. DATE

☐ RESCUE ☐ RECOVERY (Acceptable Entry Conditions > 19.5% O_2 < 10% LEL/<PEL)

PRE-ENTRY

HAZARD CONTROL

VENTILATION: ☐ Positive Pressure ☐ Exhaust ☐ Local Exhaust ☐ Local Supply

MECHANICAL: ☐ Block Linkage ☐ Disconnect ☐ None ELECTRICAL: ☐ Lockout ☐ Tagout ☐ None

PNEUMATIC: ☐ Lockout ☐ Tagout ☐ None PIPING: ☐ Blind ☐ Disconnect ☐ None

HYDRAULIC: ☐ Lockout ☐ Tagout ☐ Bleed Lines ☐ Disconnect Lines ☐ None

EQUIPMENT REQUIRED

RESPIRATORY PROTECTION: ☐ SCBA ☐ SAR ______Ft. Airline

VENTILATION: ______Fans ______Ft. Duct ______Ft. Electrical Cord ☐ Generator

LIGHTING: ☐ Caplamp ☐ Hand Light ☐ Light Sticks ☐ Cord Light ______Ft. Electrical Cord

(All equipment should be explosion-proof and equipped with GFCI)

ENTRY AND EXTRIFICATION: ☐ Tripod ☐ Davit ☐ Winch ☐ Rope M/A ☐ Belay Line ☐ Harness

VICTIM PACKAGING: ☐ Backboard ☐ Half Back ☐ SKED ☐ Litter ☐ Harness

COMMUNICATIONS: ☐ Visual ☐ Hardline ☐ Radio

ENTRY

Entry Team 1	Phone No./Radio Call Sign:
Backup Team	
Entry Team 2	
Attendant	

Atmospheric Monitoring Required:

☐ Continuosly Record on log every Min.

TERM.

Entry Terminated	Time	Date
Rescue Group/Entry Supervisor	Print	Signature

Provided by CMC Rescue, Inc. Copies of this form can be downloaded at cmcrescue.com/forms.

Confined Space Tactical Log

DATE	INCIDENT
LOCATION	

TIME	ATMOSPHERIC MONITORING				ACTION TAKEN
	LOCATION	OXYGEN	% LEL	TOX	

Provided by CMC Rescue, Inc. Copies of this form can be downloaded at cmcrescue.com/forms.

Confined Space Air Supply Log

DATE INCIDENT

LOCATION

TYPE OF BOTTLES # OF BOTTLES # OF MANIFOLDS

Entry Team 1 Air Lines		**Entry Team 2** Air Lines		**Backup Team** Air Lines	
TIME	ACTION TAKEN	TIME	ACTION TAKEN	TIME	ACTION TAKEN

AIR SUPPLY OFFICER:

Provided by CMC Rescue, Inc. Copies of this form can be downloaded at cmcrescue.com/forms.

Confined Space Rescue Incident Command System

INCIDENT DATE LOCATION

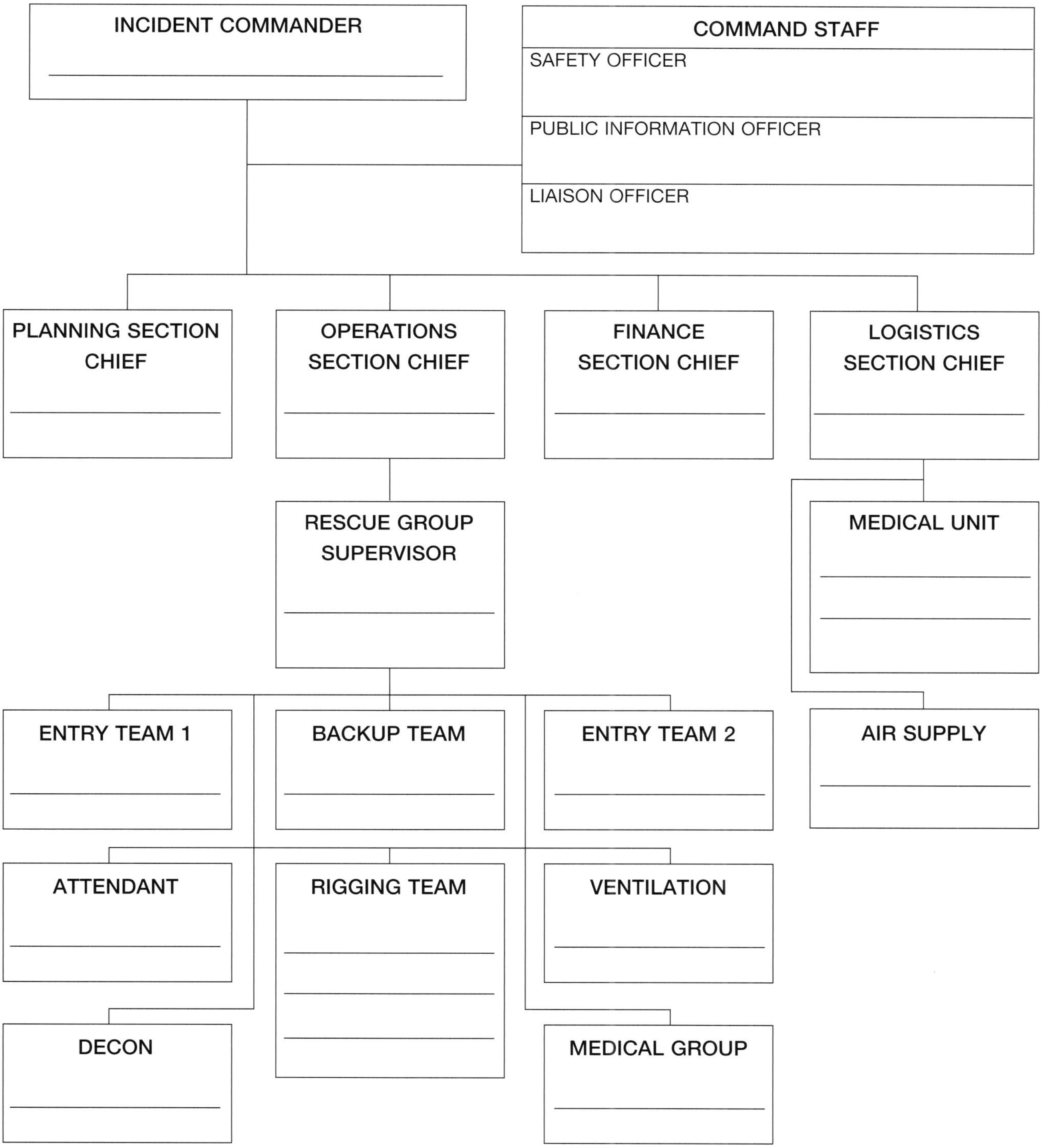

Confined Space Rescue Permit (Page 1 of 4)

This permit shall be completed in its entirety, remaining at the rescue site for the duration of the rescue operation and kept on file for one year following the event. Shading denotes an operational priority or mandatory component.

INCIDENT NUMBER	INCIDENT NAME	DATE/TIME

INCIDENT LOCATION	
RESCUE START DATE AND TIME	RESCUE END DATE AND TIME
DESCRIPTION/USE OF CONFINED SPACE	FACILITY CONTACT
SPECIAL POTENTIAL HAZARDS	

ICS Assignments

RESCUE GROUP SUPERVISOR	ATTENDANT
AUTHORIZED ENTRANT #1	BACKUP ENTRANT #1
AUTHORIZED ENTRANT #2	BACKUP ENTRANT #2
AIR MONITORING	SAFETY OFFICER
AIR SUPPLY	RIGGERS

Pre-Entry Checklist

- ☐ Operations Perimeter Setup
- ☐ Atmospheric Monitoring
- ☐ Ventilation
- ☐ Eliminate Ignition Sources
- ☐ Confirm Lockout/Tagout
- ☐ Provide Lighting
- ☐ Respiratory Protection
- ☐ Protective Clothing
- ☐ Communications
- ☐ Pre-Entry Briefing

Communication Plan

- ☐ Visual/Hand Signals
- ☐ Voice
- ☐ Radio (Intrinsically Safe)
- ☐ Rope Signals (OATH)
- ☐ Hardwire

* Identify Backup Communication Plan

Ventilation Plan

- ☐ Natural
- ☐ Forced Exhaust
- ☐ Forced Supply
- ☐ Microatmosphere
- ☐ Other:

Provided by CMC Rescue, Inc. Copies of this form can be downloaded at cmcrescue.com/forms.

Confined Space Rescue Permit (Page 2 of 4)

ENTRANT	ENTRY TIME	SCBA PRESSURE	SAR PRESSURE	EXIT TIME

GAS	PHYSICAL CHARACTERISTICS	FLAMMABILITY LEL	TOXICITY	SYMPTOMS
Carbon Monoxide CO	Colorless Odorless	12.5% 125,000 ppm	IDLH 1,200 ppm	Headache, Nausea, Dizziness, Tachypnea
Carbon Dioxide CO_2	Colorless Odorless	Non-Flammable	IDLH 40,000 ppm	Headache, Dizziness, Restlessness, Sweat, Dyspnea
Methane CH_4	Colorless Odorless	5%		
Hydrogen Sulfide H_2S	Colorless Rotten-Egg Odor	4% 40,000 ppm	IDLH 100 ppm	Eye Irritation, Respiratory Irritation, Headache
Sulfur Dioxide SO_2	Colorless Suffocating Odor	Non-Flammable	IDLH 100 ppm	Eye, Nose, Throat Irritation, Coughing, Skin Burns
Nitrogen Dioxide NO_2	Yellow/Brown Pungent Odor	Non-Flammable	IDLH 20 ppm	Cough, Frothy Sputum, Eye Irritation

Notes:

Confined Space Rescue Permit (Page 3 of 4)

Prior to entry a pre-entry briefing shall be performed with all key personnel, this includes, but is not limited to:

☐ ENTRANT	☐ ATTENDANT
☐ BACKUP ENTRANT	☐ RESCUE GROUP SUPERVISOR

The following information will be reviewed prior to entry:

- ☐ The hazards that may be encountered specific to this entry. (Atmospheric, Engulfment, Mechanical, Physical, Corrosive, Biological)
- ☐ The primary and backup communications plan.
- ☐ A confirmation that the entrant has all equipment needed to perform a successful entry and is trained on all of the equipment. (PPE, Respiratory Equipment, Communication, Rigging, Patient Packaging)
- ☐ A review of any potential self rescue plans if possible.

ENTRY AUTHORIZED

Rescue Group Supervisor:

Signature: Date/Time:

Notes:

ENTRY CANCELLED

Rescue Group Supervisor:

Signature: Date/Time:

Provided by CMC Rescue, Inc. Copies of this form can be downloaded at cmcrescue.com/forms.

Confined Space Rescue Permit

Atmospheric Monitoring Results

DATE	TIME	LEVEL	% OXYGEN	% LEL	(PPM) H_2S	(PPM) CO	INITIAL

Name (print):	Signature:	Date/Time:

Provided by CMC Rescue, Inc. Copies of this form can be downloaded at cmcrescue.com/forms.

Rescue Team or Rescue Service Evaluation Criteria

1.This appendix provides guidance to employers in choosing an appropriate rescue service. It contains criteria that may be used to evaluate the capabilities both of prospective and current rescue teams. Before a rescue team can be trained or chosen, however, a satisfactory permit program, including an analysis of all permit-required confined spaces to identify all potential hazards in those spaces, must be completed. OSHA believes that compliance with all the provisions of §1910.146 will enable employers to conduct permit space operations without recourse to rescue services in nearly all cases. However, experience indicates that circumstances will arise where entrants will need to be rescued from permit spaces. It is therefore important for employers to select rescue services or teams, either on-site or off-site, that are equipped and capable of minimizing harm to both entrants and rescuers if they need arises.

2. For all rescue teams or services, the employer's evaluation should consist of two components: an initial evaluation in which employers decide whether a potential rescue service or team is adequately trained and equipped to perform permit space rescues of the kind needed at the facility and whether such rescuers can respond in a timely manner, and a performance evaluation in which employers measure the performance of the team or service during an actual or practice rescue. For example, based on the initial evaluation, an employer may determine that maintaining an on-site rescue team will be more expensive than obtaining the services of an off-site team, without being significantly more effective, and decide to hire a rescue service. During a performance evaluation, the employer could decide, after observing the rescue service perform a practice rescue, that the service's training or preparedness was not adequate to effect a timely or effective rescue at his or her facility and decide to select another rescue service or to form an internal rescue team.

A. Initial Evaluation

I. The employer should meet with the prospective rescue service to facilitate the evaluations required by §1910.146(k)(l)(i) and §1910.146(k)(l)(ii). At a a minimum, if an off-site rescue service is being considered, the employer must contact the service to plan and coordinate the evaluations required by the standard. Merely posting the service's number or planning to rely or the 911 emergency phone number to obtain these services at the time of a permit space emergency would not comply with paragraph (k)(l) of the standard.

II. The capabilities required of a rescue service vary with the type of permit spaces from which rescue may be necessary and the hazards likely to be encountered in those spaces. Answering the questions below will assist employers in determining whether the rescue service is capable of performing rescues in the permit spaces present at the employer's workplace.

1. What are the needs of the employer with regard to response time (time for the rescue service to receive notification, arrive at the scene and set up and be ready for entry)? For example, if entry is to be made into an IDLH atmosphere, or into a space that can quickly develop an IDLH atmosphere (if ventilation fails or for other reasons), the rescue team or service would need to be standing by at the permit space. On the other hand, if the danger to entrants is restricted to mechanical hazards that would cause injuries (e.g., broken bones, abrasions) a response time of 10 or 15 minutes might be adequate.

2. How quickly can the rescue team or service get from its location to the permit spaces from which rescue may be necessary? Relevant factors to consider include the location of the rescue team or service relative to the employer's workplace, the quality of roads and highways to be traveled, potential bottlenecks or traffic congestion that might be encountered in transit, the reliability of the rescuer's vehicles and the training and skill of its drivers.

3. What is the availability of the rescue service? Is it unavailable at certain times of the day or in certain situations? What is the likelihood that key personnel of the rescue service might be unavailable at times? If the rescue service becomes unavailable while an entry is underway, does it have the capability of notifying the employer, so that the employer can instruct the attendant to abort the entry immediately?

4. Does the rescue service meet all the requirements of paragraph (k)(2) of the standard? If not, has it developed a plan that will enable it to meet those requirements in the future? If so, how soon can the plan be implemented?

5. For off-site services, is the service willing to perform rescues at the employer's workplace? (An employer may not rely on a rescuer who declines, for whatever reason, to provide rescue services.)

6. Is an adequate method for communications between the attendant, employer and prospective rescuer available, so that a rescue request can be transmitted to the rescuer without delay? How soon after notification can a prospective rescuer dispatch a rescue team to the entry site?

Rescue Team or Rescue Service Evaluation Criteria (Continued)

7. For rescues into spaces that may pose significant atmospheric hazards and from which rescue entry, patient packaging and retrieval cannot be safely accomplished in a relatively short time (15-20 minutes), employers should consider using air-line respirators (with escape bottles) for the rescuers and to supply rescue air to the patient. If the employer decides to use SCBA, does the prospective rescue service have an ample supply of replacement cylinders and procedures for rescuers to enter and exit (or be retrieved) well within the SCBA's air supply limits?
8. If the space has a vertical entry of five feet in depth, can the prospective rescue service properly perform entry rescues? Does the service have the technical knowledge and equipment to perform rope work or elevated rescue, if needed?
9. Does the rescue service have the necessary skills in medical evaluation, patient packaging and emergency response.
10. Does the rescue service have the necessary equipment to perform rescues, or must the equipment be provided by the employer or another source?

B. Performance Evaluation

Rescue services are required by paragraph (k)(2)(iv) of the standard to practice rescues at least once every 12 months, provided that the team or service has not successfully performed a permit space rescue within that time. As part of each practice session, the service should perform a critique of the practice rescue, or have another qualified party perform the critique, so that deficiencies in procedures, equipment, training or number of personnel can be identified. The critque should be given to the employer to enable it to determine whether the rescue service can quickly be upgraded to meet the employer's rescue needs or whether another service must be selected. The following questions will assist employers and rescue teams and services evaluate their performance.

1. Have all members of the service been trained as permit space entrants, at a minimum, including training in the potential hazards of all permit spaces, or of representative permit spaces from which rescue may be needed? Can team members recognize the signs, symptoms and consequences of exposure to any hazardous atmospheres that may be present in those permit spaces?

2. Is every team member provided with, and properly trained in, the use and need for PPE, such as SCBA or fall arrest equipment, which may be required to perform permit space rescues in the facility? Is every team member properly trained to perform his or her functions and make rescues and to use any rescue equipment, such as ropes and backboards, that may be needed in a rescue attempt?

3. Are team members trained in the first aid and medical skills needed to treat victims overcome or injured by the type of hazards that may be encountered in the permit spaces at the facility?

4. Do all team members perform their functions safely and efficiently? Do rescue service personnel focus on their own safety before considering the safety of the victim?

5. If necessary, can the rescue service properly test the atmosphere to determine if it is IDLH?

6. Can the rescue personnel identify information pertinent to the rescue from entry permits, hot work permits and MSDS?

7. Has the rescue service been informed of any hazards to personnel that may arise from outside the space, such as those that may be caused by future work near the space?

8. If necessary, can the rescue service properly package and retrieve victims from a permit space that has a limited size opening (less than 24 inches [60.9cm] in diameter), limited internal space or internal obstacles or hazards?

9. If necessary, can the rescue service safely perform an elevated (high angle) rescue?

10. Does the rescue service have a plan for each of the kinds of permit space rescue operations at the facility? Is the plan adequate for all types of rescue operations that may be needed at the facility? Teams may practice in representative spaces or in spaces that are worst case or most restrictive with respect to internal configuration, elevation and portal size. The following characteristics of a practice space should be considered when deciding whether a space is truly representative of an actual permit space:

    (1) Internal Configuration

    a. Open—there are no obstacles, barriers or obstructions within the space. One example is a water tank.

    b. Obstructed—the permit space contains some type of obstruction that a rescuer would need to maneuver around. An example would be a baffle or mixing blade.

Rescue Team or Rescue Service Evaluation Criteria (Continued)

Large equipment, such as a ladder or scaffold, brought into a space for work purposes would be considered an obstruction if the positioning or size of the equipment would make rescue more difficult.

(2) Elevation

a. Elevated—a permit space where the entrance portal or opening is above grade by four feet or more. This type of space usually requires knowledge of high angle rescue procedures because of the difficulty in packaging and transporting a patient to the ground from the portal.

b. Non-elevated—a permit space with the entrance portal located less than four feet above grade. This type of space will allow the rescue team to transport an injured employee normally.

(3) Portal Size

a. Restricted—a portal of 24 inches or less in the least dimension. Portals of this size are too small to allow a rescuer to simply enter the space while using SCBA. The portal size is also too small to allow normal spinal immobilization of injured employee.

b. Unrestricted—a portal of greater than 24 inches in the least dimension. These portals allow relatively free movement into and out of the permit space.

(4) Space Access

a. Horizontal—the portal is located on the side of the permit space. Use of retrieval lines could be difficult.

b. Vertical—the portal is located on the top of the permit space, so that rescuers must climb down, or the bottom of the permit space, so that rescuers must climb up to enter the space. Vertical portals may require knowledge of rope techniques or special patient packaging to safely retrieve a downed entrant.

[63 FR 66039, Dec 1, 1998]

GLOSSARY

—3—

CMC 3D™—The 3D is a compact and lightweight locking-bar descender that features two sliding bars for friction control and efficiency. The descender can be loaded or unloaded while connected to your harness for rappelling, or connected to an anchor for one- or two-person rescue lowers. UL Classified to NFPA 1983 - General Use and Technical Use.

—A—

Accessory Cord

Acceptable Entry Conditions—The conditions that must exist in a permit space to allow entry and to ensure that employees involved with a permit-required confined space entry can safely enter into and work within the space.

Access—The passage through or the approach to a confined space portal.

Accessory Cord—A generic name for 3-mm to 9-mm kernmantle cord usually manufactured to European CE standards. Sometimes called Prusik cord, accessory cord is used for Prusik loops, tag lines and various rigging and lashing applications.

Accuracy—Exactness or correctness. In confined space rescue we want our monitors to be accurate or have no errors.

A.C.G.I.H.—American Conference of Governmental Industrial Hygienists is the organization that establishes Threshold Limit Values. (TLV®).

Action Limit—The highest percentage of combustible gas detected by a combustible gas detector, which is considered as the point at which people inside should leave the area. Generally this is defined as 10% of the LFL. For rescue purposes, it may rise to 25% of the LFL, if the identity of the flammable gas is known, the characteristics of the calibration gas for the combustible gas detector are known, and a relative response can be established between the gas being detected and the combustible gas detector.

Action Plan—See incident action plan.

Air Mover—A device used to force, draw or exhaust gases through a specific assembly in order to move them from one place to another.

Air-Purifying Respirator—A device designed to protect the wearer from inhalation of harmful dusts, mists, fumes, vapors or gases by removing contaminants from the ambient air by way of a filter or chemical sorbent.

Air Supply Group Supervisor—The person within the incident command structure responsible for evaluating and coordinating fresh-air breathing apparatus supply resources on the scene of the emergency.

Alarm Settings—The set points in a monitoring device at which it indicates the atmosphere has reached a hazardous condition. For example, the alarm setting for low oxygen is 19.5%.

Anchor—A generic term for a combination of anchor points, rope, web and hardware that creates a secure attachment point to an immovable object, such as a rock, tree, structural element or vehicle.

Anchor Point—The object that web or rope is attached to or tied around. This could be a boulder, tree, piton, fire truck, I-beam or even a small building.

Anchor Strap

Anchor Strap—A manufactured component for wrapping an anchor point; usually made of web with the ends terminating in D-rings or sewn loops.

Anchor System—A group of anchor points connected together to form an anchor. If a single anchor point should fail, the system will remain intact, allowing the other anchor points to support the load.

ANSI—The American National Standards Institute oversees committees that develop standards, which are often referenced by OSHA.

Aramid—A manufactured fiber in which the fiber-forming substance is a long-chain synthetic polyamide in which as least 85% of the amide linkages are attached directly to the aromatic rings. Common brand names of aramid fibers include Nomex®, Kevlar®, Twaron® and Technora®.

Ascender

Ascenders—Mechanical devices used to grip a rope to allow the user to climb. Most designs use a cam that slides up the rope, then grips when the ascender is pulled downward. Available in handled and non-handled versions, ascenders are often used in other applications where gripping the rope is required, such as haul system ratchets or to grip a rope for pulling.

Assessment—Also called size up, the initial investigation of the incident by the first arriving personnel to determine scene hazards, number of subjects and their condition, resource needs and other information to develop the initial rescue plan.

ASTM International—A not-for-profit organization which provides a forum for the voluntary development of consensus standards for materials, products, systems and services (formerly known as American Society for Testing and Materials).

Atmospheric Hazards—Conditions present in an atmosphere that can be toxic, flammable, oxygen deficient, oxygen enriched or that obscure visibility.

Atmospheric Monitoring—The use of specialized equipment to assess the atmospheric conditions within a given area. The assessment may include (but is not limited to) oxygen content, flammability, toxicity and ambient temperature.

Attendant—An individual stationed outside one or more confined spaces who monitors the authorized entrants and who performs all attendant's duties assigned in the employer's permit space program.

Authorized Entrant—An employee who is authorized by the employer to enter a confined space.

Awareness Level—The most basic of the three levels of operational capability. Awareness level personnel and organizations are trained to recognize a technical rescue incident, know who to respond to the incident and initiate the incident command structure.

—B—

Backing Up—A redundant system of equal strength to the primary system and installed in case of a primary system failure. Typically anchors or anchor systems are backed up.

Basket Hitch

Basket Hitch—A sling around an object (or an anchor point) where the two ends come together to be connected.

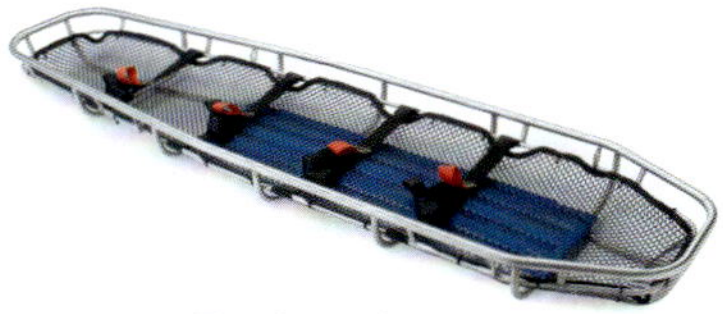

Basket Litter

Basket Litter—A patient transport device that consists of a metal frame, a surface for the patient and a protective mesh attached to the frame. Often called a Stokes litter after the person who designed the original military versions. Litter is used interchangeably with stretcher.

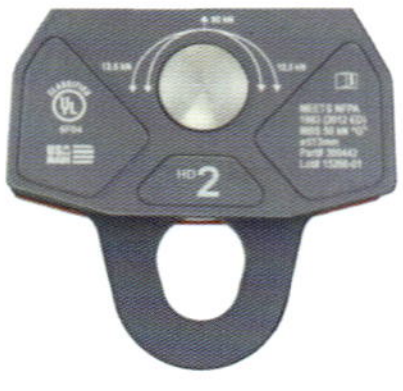

Becket

Becket—A hole at the bottom of the extended center plate of a double pulley for attaching a rope either directly or using a carabiner or quick link.

Belay—A climbing term for a safety line. A system to provide fall protection or fall arrest for yourself or to another person.

Belay Line—A line designated to provide fall protection. In a two-line system, the main line controls the movement of the load, while the belay line catches the load if a main line failure occurs.

Bend—A knot that connects the ends of two lengths of rope or web together.

BFR—A slang term used by rescuers to describe a very, very secure anchor point. Originally a BIG rock, but could refer to any immovable object, such as large trees, fire trucks, water tanks or bridge abutments. Immovable is the key element, which usually, but not always, equates to size.

Bight

Bight—Formed when the rope is doubled back but does not cross itself.

Biological Hazards—Microbial agents presenting a risk or potential risk to the well-being of humans through inhalation, ingestion, skin absorption or injection.

Blank Flanges—A plate bolted to a flange that has no opening in it and is meant to block the flow of a product past that point in a pipe.

Blanking/Blinding—Inserting a solid barrier, such as a spectacle blind or skillet blind, across the open end of a pipe or in between two flanges leading into or out of the confined space and securing the barrier in such a way to prevent leakage of material.

Block and Tackle—A simple mechanical advantage system with a single rope reeved through a series of pulleys.

Blocks—An energy-isolating device meant to stop or obstruct the flow of hazardous energy or products.

Body Belt—A belt that fits around the waist and has a load-bearing attachment point. Some standards now limit the use of a body belt to travel restraint, positioning on a ladder or emergency escape.

Bolted Slip Blind—A blinding device that is meant to be bolted directly between flanges to stop or obstruct the flow of product.

Bombproof—An adjective for an anchor or anchor point that the rescuer is absolutely certain will support the expected and any unexpected loads placed on the rope system.

Brake Bar Rack

Bottom Belay—A belay method in which the rappel rope is tended at the bottom by the belayer. By pulling down on the rope, the belayer can slow or stop the descent. A bottom belay protects against loss of control but does not protect against a failure of the anchor or rappel line or a failure to rig the descender onto the rappel line.

Brake Bar Rack—A lowering or descent control device utilizing a series of bars oriented in a ladder like pattern on a J- or U-shaped steel frame.

Breathing Air System—Equipment used to provide respiratory protection to personnel working within a space that contains an actual or potential respiratory hazard.

Bridle

Bridle—Along with "spider" another name for a litter or stretcher harness.

Buddy System—A system of organizing employees into work groups in such a manner that each employee of the work group is designated to be observed by at least one other employee in the work group. The purpose of the buddy system is to provide rapid assistance to employees in the event of an emergency.

Bump Test—A field check of a monitoring instrument to verify that the monitor is responding and that the alarms are working. Does not replace the need to calibrate the monitor.

—C—

Cal/OSHA—The State of California agency that provides occupational safety and health administration.

Calibrated—The status of a measuring instrument after its accuracy has been checked or corrected.

Calibration—A laboratory or bench test resetting of the instrument's zero, span and alarm points according to manufacturer's specifications. Calibration is performed by a factory authorized service center or trained technician.

Cam—The moving element of an ascender or rope grab that is forced toward the frame or shell to grip the rope when a load is placed on the device. Also a slang term for a rope grab used in a rope rescue mechanical advantage system.

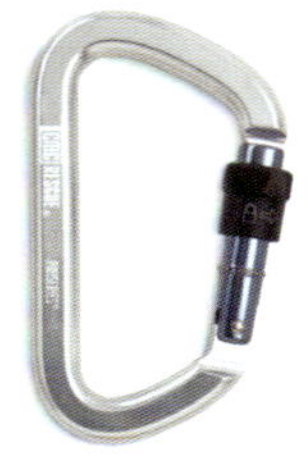
Carabiner

Carabiner—A D- or oval-shaped load-bearing connector with a self-closing gate used to join life safety rope and system components.

Chain Stitch—A type of construction of tubular web in which two layers of flat web are stitched together on the edge using a lock stitch.

Class II Harness—The NFPA Standard 1983 designation for a harness that fits around the waist and thighs, designed to be used for rescue, with a design load of 600 lbf.

Class III Harness—The NFPA Standard 1983 designation for a harness that fits around the shoulders, waist and thighs, designed to be used for rescue, with a design load of 600 lbf.

Colormetric Tube—See detector tube.

Combustible Gas Indicator—A device that detects and measures the presence of a flammable gas based on how close the gas is to the lower flammable limit of the calibration gas.

Command Post (CP)—The location at which the primary command functions are executed. The CP may be colocated with the incident base or other inci-dent facilities.

Complex System—A rope rescue mechanical advantage system that is neither a simple system nor a compound system.

Compound System—A rope rescue mechanical advantage system in which a simple system is pulling on another simple system.

Confined Space—A space that is large enough and so configured that an employee can bodily enter and perform assigned work, has limited or restricted means for entry or exit and is not designed for continuous employee occupancy. Tanks, vessels, silos, storage bins, hoppers, vaults and pits are spaces that may have limited means of entry are examples of confined spaces.

Confined Space Rescue—The removal of a trapped, sick or injured person from a confined space or permit-required confined space.

Confined Space Supervisor—See entry supervisor.

Contamination Reduction (Warm) Zone—The area between the Exclusion (Hot) Zone and the Support (Cold) Zone, where PPE is removed and decontamination occurs.

Contents—For confined space rescues, the materials within the confined space. The contents may be either gas, liquid or solid and may or may not contribute to the limiting factors affecting the confined space emergency.

Controlled Breathing—A breathing technique that creates efficient use of existing fresh-air supply when performing normal work functions and wearing fresh-air breathing apparatus.

Contingent Anchor—A lowering (or raising) system incorporated into the anchor to allow a person stuck on line to be quickly rescued by lowering or raising the line.

Corrosive—An acidic or basic material that can damage human skin or rapidly corrode unprotected metal.

Counter Balance—A rope rescue system that uses a change of direction pulley and an offsetting load. The offsetting load traveling down reduces the force that the haul team needs to exert to lift the primary load.

Critical Incident Stress Debriefing or Critical Incident Stress Management—A pre-planned process for identifying and treating psychological trauma resulting from participation in an incident.

Critique—An honest, searching evaluation of the performance of the rescue team for the purpose of improving the service provided to the community. The inability to critique suggests a dangerous overconfidence based on intentional ignorance.

—D—

Debriefing—A session following an incident to critically analyze all conditions and actions taken in the management of that emergency. The purpose is to determine effectiveness of operations and how to improve them when managing similar incidents in the future.

Decibel—A unit for measuring the relative loudness of sound.

Decontamination—To make safe by eliminating toxic or otherwise harmful substances, such as noxious chemicals or radioactive material.

Defensive Actions—Actions taken in which there is no intentional contact with the hazards of a confined space. Rescue from outside of a confined space by retrieving the subject with in-place retrieval equipment is one example.

Degradation—The reduction of protective properties of chemical protective clothing by mechanical, thermal or chemical means with a loss of integrity of the garment.

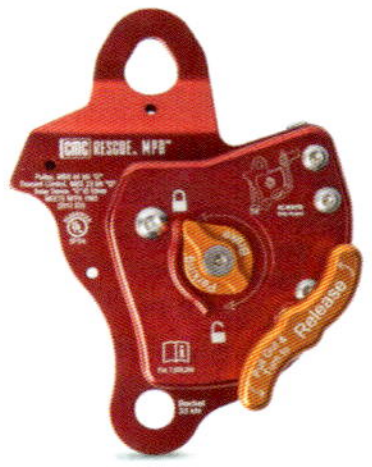

Descent Control Device

Descent Control Device (DCD)—A friction device attached to a rope to control the lowering of a load; used for rappels or lowering systems.

Destructive Test—A test method in which the test sample is taken to failure.

Detection Range—The area between the upper and lower limits in which a monitor will detect and measure accurately.

Detector Tube—A device consisting of a glass tube filled with a solid material that reacts chemically with an air contaminant drawn through it, usually with a hand pump, resulting in a color change the length of which is proportional to the contaminant concentration.

Direct-Reading Instruments—Detection and monitoring instruments that give a reading based on a graduated scale and do not require any calculations to determine a result.

Directional—Using a pulley to bend a line to turn corners in the ascent or descent or to place the haul or belay system into a more efficient direction of travel.

Disconnect Switches—An electrical switch designed to isolate the electrical source from the equipment that it powers by disconnecting the power supply from the equipment.

Distributing Anchor—See load distributing anchor.

Double Block and Bleed—A method used to isolate a confined space from a line, duct or pipe by physically closing two main valves on a piping system and opening a vented-to-atmosphere valve between them.

Dressing—Eliminating open loops and twisted or crossed rope segments to cause a knot to tighten in a neat orderly fashion for the purpose of maximizing knot strength.

Dusts—Fine particles of material suspended in the atmosphere.

Dynamic Load—A force rapidly applied to a system, such as when a fall is arrested by a safety or belay line; sometimes referred to as an impact force or impact load.

Dynamic Rope—A rope designed to minimize the impact force of a fall through elongation, such as a recreational climbing rope. Low-stretch ropes are defined as less than 10% elongation at 10% of breaking strength, suggesting that a dynamic rope would be a rope with greater than 10% elongation.

Dyspnea—Difficult or labored breathing.

—E—

Edgeman—Also called an edge tender, this is the person positioned at the edge of a vertical space who can assist with moving the stretcher over the edge, relaying communications as needed and tending edge protection devices. In a high-angle rescue, the edgeman is capable of descending and ascending. Otherwise, if a fall hazard exists, the edgeman is provided with a travel restraint line.

Elevated Rescue—A rescue of a subject from heights where normal means of egress are not available.

Elongation—The amount that a rope stretches at a specified load, usually reported as a percentage of the rope's length before the load is applied.

Emergency—Any occurrence (including any failure of hazard control or monitoring equipment) or event internal or external to the permit space that could endanger entrants.

Engulfment—The surrounding and effective capture of a person by a liquid or finely divided (flowable) solid substance that can be aspirated to cause death by filling or plugging the respiratory system or that can exert enough force on the body to cause death by strangulation, constriction or crushing.

Entrant—A person who enters a confined space to perform and assigned task. See also authorized entrant.

Entry—Ingress by person into a confined space, which occurs upon breaking the plane of the confined space portal with any part of the body. Entry includes all periods of time in which the confined space is occupied.

Entry Permit—The written or printed document that is provided by the employer to allow and control entry into a permit space and that contains the information specified by the governing regulation.

Entry Supervisor—The person (such as the employer, foreman or crew chief) responsible for determining if acceptable entry conditions are present at a permit space where entry is planned, for authorizing entry and overseeing entry operations and for terminating entry as required. An entry supervisor also may serve as an attendant or as an authorized entrant, as long as that person is trained and equipped as required for each role he or she fills. Also, the duties of entry supervisor may be passed from one individual to another during the course of an entry operation.

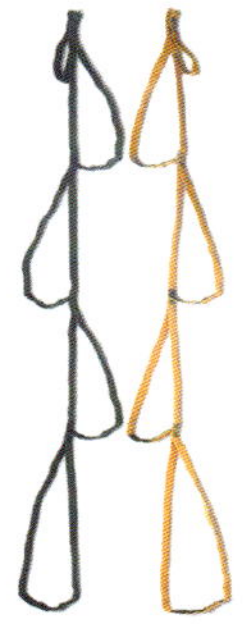

Etrier

Etrier—A set of loops or stirrups sewn or tied from webbing and used for ascending. It's a French climbing term and is pronounced *á-tray-á.*

EPA—Environmental Protection Agency. Their mission is to protect human health and the environment.

Escape Bottle—An air bottle that normally holds five to 10 minutes of air and is worn by an entrant to provide sufficient air for an emergency egress if there is a failure in the primary air supply source.

Evacuation—An unaided emergency exit out of a confined space. This action may result from the entrant's own decision or by a command from outside the space.

Exclusion (Hot) Zone—The area where hazards exist or have the potential to exist, and PPE or other safety equipment is required.

Exhaust Ventilation—Using a blower to create a negative pressure within a confined space. The hazardous atmosphere is removed through the blower and its ducting and is replaced by fresh air entering the space.

Explosive Limits—The lowest (leanest) and highest (richest) ratios of a fuel and oxygen mixture that would support ignition.

Explosive Range—The area between the LEL and UEL of a fuel and oxygen that would support ignition. Mixtures below the LEL are too lean and above the UEL are to rich to support combustion.

Exposure—The status of being near or in the vicinity of a physical hazard, which could be mechanical, temperature, elevation, biological or radiological.

—F—

Fall Arrest—Stopping the downward movement of a falling individual.

Fall Factor—The height of a fall divided by the length of the rope available to arrest the fall. Essentially the fall factor value compares the potential energy in the system with its capacity to absorb that energy. The higher the fall factor, up to a maximum of 2, the greater the severity of the fall.

Fall Protection—Equipment or procedures that prevent a person at height from falling or, should a fall occur, arrests the fall. Fall protection can be part of the structure, worn by the person or procedural policy. It includes both fall arrest and travel restraint.

Figure 8 Descender

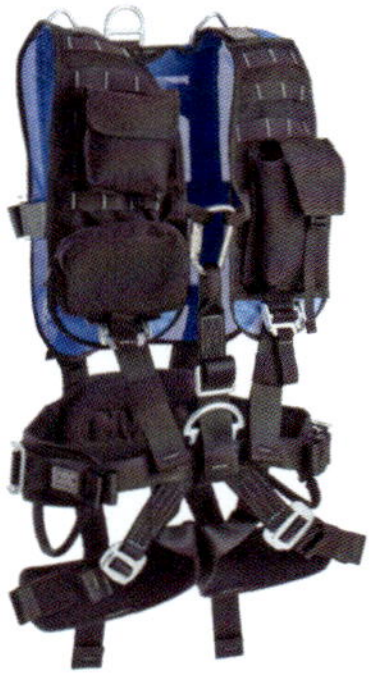

Full-Body Harness

Field Test—Verification of performance under typical conditions of use.

Figure 8 Descender—A metal plate with the approximate shape of the number eight that is used as a descent control device.

Fixed Line—A non-moving rope attached to an anchor.

Flammable Range—The area between lowest and highest concentrations of a gas that would support ignition.

Flash Point—The lowest temperature at which the vapor of a combustible liquid can be made to ignite momentarily in air.

Friction—The rubbing of one object or surface against another, which resists the motion of one relative to the other. Friction increases the work that a haul team must do to lift a load but reduces the load on the system during lowering.

Full-Body Harness—A harness that fastens about the upper thighs, pelvis, torso and shoulders.

—G—

General Use—An equipment designation in NFPA Standard 1983 Life Safety Rope and Equipment for Emergency Services. General Use has the highest performance requirements of the three designations: General, Technical and Escape.

—H—

Handline—A fixed line used to assist and to self belay an ascent or descent by providing a rope to hold onto. The rope is usually gripped by the hand but a rope grab could also be used.

Haul Cam—The rope grab used to attach a moving pulley to the rope when assembling a mechanical advantage system. Different than the ratchet cam, which connects to the anchor and holds the load while the system is being reset or modified.

Hazardous Atmosphere—An atmosphere that may expose employees to the risk of death, incapacitation, impairment of ability to self-rescue, injury or acute illness from flammable substances; combustible dust; oxygen deficiency or enrichment; toxic substances or any other atmospheric condition that is immediately dangerous to life or health.

Hazard Reduction—The elimination or mitigation of a hazard within a specific area.

Hazardous Material—A solid, liquid or gas that when released is capable of causing harm to people, property or the environment.

HAZWOPER—OSHA's Hazardous Waste Operations and Emergency Response Standard, 29 CFR 1910.120.

High Angle—When the angle of the terrain becomes so steep that the weight of the individual or the stretcher is supported primarily by the rope rather than by the feet.

High Point Anchor—An attachment point for a rope rescue system component elevated above the working level of the system.

Hitch—A knot that attaches a rope to another object, including another rope, if the host rope is not part of the knot. In most cases a hitch can be slid along the object and will fall apart if the object is removed.

Horizontal Rescue—Rescue procedures where the individual's weight is supported by a horizontal surface, such as a floor or horizontal surface inside the space.

Horizontal Litter—When the litter or stretcher is positioned perpendicular to the rope (or the fall line). A litter harness or spider is used for support.

Hot Work—Work that produces a potential source of ignition, such as sparks, flames, heat or arcs.

Hypothermia—A decrease in body core temperature.

—I—

Incident Command System (ICS)—The nationally used, standardized on-scene emergency management concept specifically designed to allow its user(s) to adopt an integrated organizational structure equal to the complexity and demands of single or multiple incidents without being hindered by jurisdictional boundaries. ICS is the combination of facilities, equipment, personnel, procedures and communications operating within a common organizational structure with responsibility for the management of resources to effectively accomplish stated objectives pertinent to an incident.

IDLH—Immediately Dangerous to Life or Health means any condition that poses an immediate or delayed threat to life, that would cause irreversible adverse health effects or that would interfere with an entrant's ability to escape unaided from a permit space.

Impact Load—See dynamic load.

Incident Action Plan—the plan developed at the field response level which contains objectives reflecting the overall incident strategy and specific tactical actions and supporting information for the next operational period.

Incident Commander—The individual responsible for the command of all functions at the field response level.

Independent Belay—A belay system that is entirely separate from the main line system. It provides the highest level of protection but requires a second anchor, rope, hardware and operator.

Inerting—The displacement of the atmosphere in a permit space by a noncombustible gas (such as nitrogen) to such an extent that the resulting atmosphere is noncombustible. Inerting creates an IDLH oxygen-deficient atmosphere.

Internal Configuration—The arrangement of partitions, pipes, baffles, mixers, augers or other devices that could restrict access or movement in a space.

Intrinsically Safe—Equipment and associated wiring in which any spark or thermal effect, produced either normally or in specified fault conditions, is incapable, under certain prescribed test conditions, of causing ignition of a mixture of flammable or combustible material in air in its most easily ignitable concentration.

Irritant—Substances harmful to persons but that do not cause long-term effects.

Isolation—The process by which a confined space is removed from service and completely protected against the release of energy and material into the space by such means as blanking or blinding, misaligning or removing sections of lines, pipes or ducts, a double block-and-bleed system; lockout or tagout of all sources of energy or blocking or disconnecting all mechanical linkages.

—K—

Kernmantle—German for core (kern) and sheath (mantle) and refers to a type of rope construction in which a load-bearing core is covered by a protective sheath.

Kevlar®—A brand of aramid fibers manufactured by DuPont and used to manufacture rope, cord, web and fabrics. Kevlar brand fibers have a high strength to weigh ratio, very low elongation and a high heat and flame resistance.

Knot—A fastening made by tying rope or cord. Knot is generic but also specifically applies to a knot that is not a hitch or a bend.

—L—

Laid Rope—A type of rope construction in which three or more strands are twisted together.

Latch—A fastening device that consists of a bar that falls into a notch to prevent opening or operation of the object it secures.

Lead Time—The time between the decision to start a task and the completion of the preparation necessary to start the work; setup time.

LEL/LFL—Lower Explosive Limit/Lower Flammable Limit is the leanest mixture of fuel and oxygen that will permit ignition.

Level A—PPE to be selected when the greatest level of skin, respiratory and eye protection is required. It usually requires a positive-pressure, full face piece SCBA or positive-pressure SAR with escape SCBA and a totally encapsulating chemical-protective suit. Level A protection should be worn when the hazardous substance has been identified and requires the highest level of protection for skin, eyes and the respiratory system, based on either the measured (or potential for) high concentration of atmospheric vapors, gases or particulates; or the site operations and work functions involve a high potential for splash, immersion or exposure to unexpected vapors, gases or particulates of materials that are harmful to skin or capable of being absorbed through the skin; substances with a high degree of hazard to the skin are known or suspected to be present and skin contact is possible; or operations must be conducted in confined, poorly ventilated areas and the absence of conditions requiring Level A have not yet been determined.

Level B—PPE when the highest level of respiratory protection is necessary but a lesser level of skin protection is needed. Level B equipment usually requires a positive-pressure, full-face-piece SCBA or positive-pressure SAR with escape SCBA and hooded chemical-resistant clothing. Level B protection should be used when the type and atmospheric concentration of substances has been identified and requires a high level of respiratory protection, but less skin protection; the atmosphere contains less than 19.5% oxygen; or the presence of incompletely identified vapors or gases is indicated by a direct-reading organic vapor detection instrument, but vapors and gases are not suspected of containing high levels of chemicals harmful to skin or capable of being absorbed through the skin.

Level C—Level C protection should be used when the atmospheric contaminants, liquid splashes or other direct contact will not adversely affect or be absorbed through any exposed skin; the types of air contaminants have been identified, concentrations measured and an air-purifying respirator is available that can remove the contaminants; and all criteria for the use of air-purifying respirators are met. Level C equipment may include a full-face or half-mask, air-purifying respirators and chemical-resistant clothing.

Level D—Level D protection should be used when the atmosphere contains no known hazard and the work functions preclude splashes, immersion or the potential for unexpected inhalation of or contact with hazardous levels of any chemicals. Protective eyewear and a work uniform affording minimal protection is used for nuisance contamination only.

Liaison Officer—A member of the command staff at the field level responsible for coordinating with representatives from cooperating and assisting agencies. At EOC levels, the function may be done by a coordinator and/or within a section or branch reporting directly to the EOC director.

Line—Another term for a rope, usually refers to a rope that is in use.

Line Transfer—A rope rescue skill for transferring an individual on a line to a rescue line or system.

Line Valves—Valves in pipes entering a confined space that can be used to prevent the flow of material into the space.

Litter Attendant—Person or persons assigned to manage the litter or stretcher during movement. Litter tender is becoming the preferred term since the attendant is a unique position in confined space rescue (see attendant). The litter tender guides the litter around obstacles, tends to the patient and communicates with the rescuers operating the rope systems.

Load—A generic term for the force generated by everything that is hanging on the rope at the end away from the anchor. This could include a rescuer, patient, stretcher and tender(s), everyone's gear and the system hardware. The load lifted by the haul team would also include the friction in the system.

Load-Distributing Anchor System—An anchor system that, in theory, automatically equalizes the load among two or more anchor points as the direction of pull shifts or if any anchor points should fail. Friction in the system prevents the loads on each anchor point from reaching an equal state and often a physical adjustment is required to create even a rough equality.

Load Limiter—A device designed to slip or release at a pre-determined force to prevent excessive strain on a system. One of the advantages of using a properly sized Prusik hitch for a rope grab is that it can also act as a load limiter, slipping before the load causes a system failure.

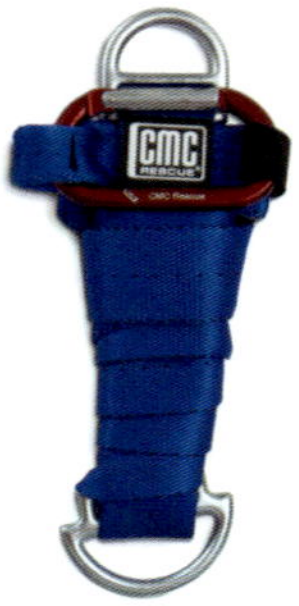

Load-Releasing Hitch

Load-Releasing Hitch—A hitch that can be untied or released while under load and then be used to transfer the load to another line in a controlled manner. Also called a load-release hitch or mariner's knot, it can be tied from cord or web. A load-release strap is a commercially sewn version.

Load-Sharing Anchor System—An anchor system that is constructed such that the load on each leg will be as equal as possible. The actual level of equality is very dependent on the ability the rigger to estimate the direction of pull and the length of each leg. Since the legs are of fixed length, a failure of an anchor point will cause a shift in the direction of pull but will not generate slack in the system as load-distributing anchor system will.

Local Exhaust—A type of ventilation system used when the contaminants are located or generated at one or more specific points. The exhaust intake is placed at or near the source or point of concentration of the contaminants. Frequently used for hot work, grinding operations or cleaning with solvents.

Lock-Off—Securing the rope around a descent control device so that the rope will not slide through when the operator lets go of the rope and the device.

Lockout/Tagout—The control or elimination of electrical and mechanical energy sources within the confined space by physically attaching locking devices and tags.

Logarithmic Scales—A measuring scale in which each integer value represents an increase by the power of 10. A common example is the pH scale where a value of 5 indicates ten times greater acidity than a value of 6.

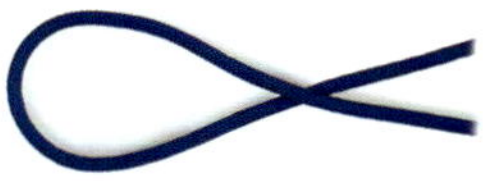
Loop

Loop—Formed when a rope is doubled back and crosses over itself.

Low Angle—The angle of terrain where the weight of the stretcher is supported primarily by the tender's legs, and the rope system is required to both facilitate movement and for fall protection. The actual angle in degrees will vary depending on the type of surface and the skill level of the tenders.

Low-Point Anchor—An attachment point for a rope rescue system component located at or below the working level of the system.

Low-Stretch Rope—Defined by the Cordage Institute as a rope having between 6% and 10% elongation at 10% of the rope's breaking strength.

—M—

Major Axis—The long axis of a carabiner in which it is intended to be loaded. It is parallel to the spine of a D-shaped carabiner.

Main Line—The rope designated as the primary line in a two-rope system, even though both ropes may have identical setups or be equally loaded. In a raising system, it will be the haul line with the mechanical advantage. In a lowering system, it controls the rate of descent.

Margin of Safety—The strength of the system minus the maximum load that will be put on the system. The system's strength is determined by the weakest component in the system and the maximum load, including friction, force multipliers and other external forces. The margin of safety is the buffer against disaster when the rescue does not go as planned.

Mechanical Advantage (M/A)—Any means of increasing an individual's ability to lift or move a load. In rope rescue, mechanical advantage almost always refers to a system using pulleys to increase the effectiveness of the haul team. M/A is referred to as the ratio between the output and the input of the system, for example a 3 to 1 M/A system.

Medical Monitoring—Evaluation of entrant's vital signs (pulse, respirations, temperature, blood pressure) before and after the entry to determine any significant effect.

Microatmosphere—The atmospheric conditions immediately surrounding the subject in a confined space rescue. Positive pressure ventilation directed at the subjects face may provide a respirable microatmosphere where the rest of the space may be hazardous.

Minor Axis—The short axis of a carabiner from the gate across to the spine.

Monitoring—The identification and quantification of hazardous contaminants in a confined space prior to and during entry.

MPD™—The CMC MPD (Multi-Purpose Device) is a pulley, rescue belay and descender all in one. Its high-efficiency pulley, with an integral rope-grab mechanism, means it can be used as a lowering device on the main line and belay line systems and be quickly changed over to a raising device without switching out or replacing hardware.

MSDS—Material Safety Data Sheets are written or printed material concerning a hazardous chemical and prepared in accordance with 29 CFR 1910.120 (g). A material's MSDS provides information on chemical properties, health and safety hazards, medical and emergency response to a spill or contact and proper disposal.

Multi-Point Anchor—An anchor system using more than one anchor point. It is used when an adequately strong single anchor point is not available.

Munter Hitch—A friction hitch tied using a carabiner that can be used to control a descent, such as for a belay or a rappel.

—N—

Needle Loom—A type of construction of tubular web in which two layers of web are stitched together on the edge using a lock stitch

Negative Pressure Ventilation—Using a blower to create a negative pressure within a confined space. The hazardous atmosphere is removed through the blower and its ducting and is replaced by fresh air entering the space. Also called exhaust ventilation.

NFPA International—A non-profit organization best known for development of industry consensus standards and codes on fire, electrical and building safety (formerly known as National Fire Protection Association).

NIOSH—National Institute for Occupational Safety and Health an agency established to help assure safe and healthful working conditions for working men and women by providing research, information, education and training in the field of occupational safety and health.

Nomex®—A brand of aramid fiber manufactured by DuPont that has an extraordinary combination of high-performance heat- and flame-resistant properties, as well as superior textile characteristics.

Non-Directional Anchor—An anchor system where the loads at each anchor point remain roughly equal when the direction of the pull shifts.

Non-Entry Rescue—A rescue procedure that does not require the rescuer to enter the space. This is usually a retrieval system operated by the attendant.

Non-Permit Confined Space—A confined space that does not contain or, with respect to atmospheric hazards, have the potential to contain any hazard capable of causing death or serious physical harm.

—O—

OATH—An a acronym for remembering non-verbal communication codes, such as rapping and tapping, light flashes, whistle blasts or rope tugs.

Off-Site Rescue Service—A rescue team not located on the employer's facility and responsible for rescue at that location.

On-Site Rescue Service—A rescue team located on the employer's facility and responsible for rescue.

Operations Level—The second level of the three levels of operational capability. Operational level personnel and organizations are trained to perform hazard identification, use equipment and perform limited technical rescue techniques, usually under the direction and control of technician level responders.

OSHA—Occupational Safety and Health Administration is part of the U.S. Department of Labor and responsible for developing and enforcing workplace safety and health regulations.

Oxygen-Deficient Atmosphere—An atmosphere containing less than 19.5% oxygen by volume.

Oxygen-Enriched Atmosphere—An atmosphere containing more than 23.5% oxygen by volume.

Oxygen Sensor—An electrochemical cell in which oxygen reacts with the electrolyte generating a current and voltage that indicates the percentage of oxygen present in the test atmosphere.

—P—

Packaging—Preparing a non-ambulatory patient for transport with considerations for the medical complaints, protection from the environment and prevention of injury during the evacuation.

PAL—Personal Audible Locator, uses an audible tone to signal that an entrant is in distress. Same as PASS.

Parts Per Million (ppm)—Used to express very low ratios, it is the number of units of a particular substance occurring in a total population of one million units. One part per million is 1/10,000 of 1%.

PASS—Personal Alert Safety System, a device that sounds an alarm when the wearer has been immobile for a specified length of time. See NFPA Standard 1982 for certification requirements. Same as PAL.

PEL—Permissible Exposure Limit means the exposure, inhalation or dermal permissible exposure limit specified in 29 CFR Part 1910, Subparts G and Z.

Performance Evaluation—Determination of whether an individual can perform a standardized skill.

Permit-Required Confined Space—A confined space that has one or more of the following characteristics: contains or has a potential to contain a hazardous atmosphere; contains a material that has the potential for engulfing an entrant; has an internal configuration such that an entrant could be trapped or asphyxiated by inwardly converging walls or by a floor which slopes downward and tapers to a smaller cross section; or contains any other recognized serious safety or health hazard. Also called a permit space.

Permit System—The employer's written procedure for preparing and issuing permits for entry and for returning the permit space to service following termination of entry.

Physical Hazards—Mechanical, electrical, chemical and thermal hazards in a confined space that are dangerous to entrants.

Pick-Off—A rescue technique in which the rescuer descends to the subject, attaches the subject to the rescuer's system and then both continue the descent to safety. Pick-offs can be done on rappel or with a lowering system. While usually used as a descent, the rescuer and subject could also be raised. Sometimes referred to as a line transfer rescue when the subject is suspended on a line.

Piggy Back System—Sometimes called a pig rig, it is an independent system that attaches to the main line.

PIO—The Public Information Officer is the individual at field or EOC level that has been delegated the authority to prepare public information releases and to interact with the media. Duties will vary depending upon the agency and ICS level.

Portable High Anchor—A manufactured device designed to support human loads such as davits, A-frames, tripods, quad pods and cantilevered devices.

Positive Pressure Ventilation (PPV)—Pushing air into the space to cause the contaminated air to exit through any available opening.

PPE—Personal protective equipment as appropriate for the identified hazards.

Pre-Tensioned Tieback—Connecting a less than satisfactory anchor located at the focal point to a stronger anchor using a mechanical advantage system to tension the rope or web connecting the two points.

Precision—The least variation from a standard.

Preplan—A written plan for a rescue response that provides the organization and initial actions for the most common incidents of the agency. It is based on past responses, hazard assessment and available resources.

Primary Anchor—The anchor point within an anchor system intended to support most of the load or potential load.

Prusik Hitch—A friction knot that can be slid along a rope, but when under load in either direction, will grab the rope.

Prusik-Minding Pulley (PMP)—A pulley in which the side plates are shaped to prevent a Prusik hitch from being pulled into the pulley and jamming. Most typically a PMP will have a square shaped bottom.

Purging—A method by which gases, vapors or other air contaminants are displaced from a confined space.

Push/Pull Ventilation—A ventilation method that uses both positive and negative methods to more rapidly or thoroughly ventilate a space.

—R—

Rappel—The controlled descent of a fixed line, most often with the use of a descent control device.

Release Knot—A type of knot that can be untied while under load. Some release knots will allow a controllable release of the load for transfer to another line.

Rescue—The location, access, stabilization and transport to safety or medical care of a subject at risk.

Rescue Group Supervisor—The individual responsible for the rescue group, which usually includes the attendant, ventilation, entry team, backup team, rigging team, decon and medical group. Reports to the operations section chief.

Rescue Load—ASTM Standard F2266-03 specifies a Type IV mass as 200 kg (440 lb.) and a Type V mass as 280 kg (617 lb.). The explanatory text describes the Type IV mass as representing a "rescue load with two rescuers" and the Type V mass as a "three-person rescue load."

Retrieval—Assisting an entrant to exit the confined space without the rescuers entering the space.

Retrieval Line—A rope secured to an anchor point or lifting device outside of a confined space and attached to a harness or wristlets and worn by entrants to a confined space.

Retrieval System—The equipment (including a retrieval line, chest or full-body harness, wristlets, if appropriate, and a lifting device or anchor) used for non-entry rescue of persons from permit spaces.

RFI—Radio Frequency Interference is electromagnetic radiation caused by a signal-generating electrical device. Radio and microwave devices are the most obvious signal generators, but any electrical device can emit RF radiation.

Rigger—The individual responsible for the coordination and evacuation of the patient and entry teams and the operation of the rope rescue or retrieval systems. Supervisor of the rigging team.

Rigging—The rope and hardware used to assemble a rope rescue system; includes anchor systems, main line and belay line. The act of assembling a rope rescue system.

Rope Grab Device—A device used to grasp a life safety rope for the purpose of supporting a load. Often includes ascenders.

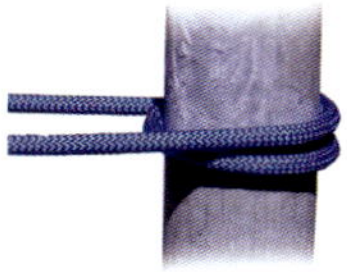
Round Turn

Round Turn—A full wrap of rope completely around and object and with both ends pointing the same direction. A 540° turn.

Running End—The end of the rope that you are working with. Also called the working end, bitter end or tag end.

—S—

Saddle Vent

Saddle Vent®—A ventilation duct used in a narrow opening and designed to allow the maximum amount of air flow with the minimum restriction to entry and egress.

Safety Factor—Margin of safety, expressed as a ratio of margin of safety divided by the maximum load.

Safety Knot—Half a double fisherman's (barrel knot) or an overhand knot tied just behind the primary knot as a backup to prevent the primary knot from working loose.

Safety Officer—A member of the command staff at the incident or within an EOC responsible for monitoring and assessing safety hazards or unsafe situations and for developing measures for ensuring personnel safety.

SAR—Supplied Air Respirator which consists of a full face mask connected by a hose to an air supply at a maximum of 300 feet away.

SCBA—Self-Contained Breathing Apparatus consisting of a full face mask, respirator and air bottle carried by the entrant. Closed-circuit rebreathers are another form of SCBA.

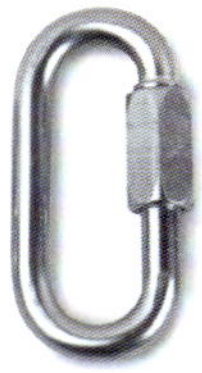

Screw Link

Screw Link—A metal connector which uses a nut for the gate and closes by screwing the nut onto the threads on the far side of the gate.

Secondary Survey—The thorough assessment of the patient by medical personnel to determine the type and extent of all injures and other adverse medical conditions.

Self Belay—A belay system operated by the person on rope. It protects against a mistake in rappelling or ascending. If it is attached to a separate rope, it also protects against a failure in the fixed line and its anchor.

Self-Equalizing Anchor—See load-distributing anchor system.

Self Rescue—The ability of the subject or entrant to exit the confined space without assistance.

Self-Retracting Lifeline—A device that connects to the fall arrest attachment point of the harness with a line that extends or retracts as the individual moves, so that a slight tension remains in the line at all times. In the case of a fall, the device provides fall arrest by locking. Also called a self-retracting lanyard.

Sensor—The element in a monitor that detects and measures the presence of a gas in the atmosphere. Sensors come in a variety of types; some are gas specific, and some are capable of measuring a wide range of contaminants.

Sheave—The wheel in a pulley. The diameter of the sheave is measured at the tread (center) of the wheel.

Shuttle Loom—A construction type of tubular webbing, also called spiral stitch.

Side Load—A condition in which the force applied to a carabiner improperly comes from the side, essentially attempting to bend the carabiner.

Side Plate—The large, flat part of a pulley that attaches to the axle and wheel assembly and has a hole at the top for the carabiner. On many rescue-type pulleys, the side plates rotate to allow the pulley to be attached without having to thread the end of the rope through.

Simple System—A rope rescue mechanical advantage system in which all moving pulleys travel in the same direction and at the same speed as the load.

Size Up—The rescue team leader's on-scene analysis of the parameters of the incident, including type of rescue involved, number of subjects, resources responding, physical and environmental hazards and the need for any additional resources, special capabilities or logistics.

SKED Stretcher®—A flexible sheet of plastic that becomes semi-rigid when folded around a patient. Manufactured by Skedco, Inc.

Snap Link—A term common in the military for a carabiner.

Spectacle Blind—A combination blind and spacer formed from one piece of metal. Frequently a permanent part of the line with the open and solid ends switched depending on whether or not the flow needs to be blocked.

Spiral Stitch—The type of construction of tubular web produced by a shuttle loom. The fibers appear to spiral around the webbing.

SPRAT—The Society of Professional Rope Access Technicians is an organization of individuals and companies in the rope access industry. Through a consensus process, SPRAT develops safe-practice standards and Rope Access Technician certification standards.

Standby Team—A trained rescue response unit located outside of the confined space and in communication with the entrants.

Static Rope—Defined by the Cordage Institute as a rope having greater than 1% and less than 6% elongation at 10% of the breaking strength.

Static Electricity—An accumulation of electric charge on an insulated body or the resultant discharge.

STEL—Short Term Exposure Limit is the concentration of material that a worker may be exposed to for a short period of time without suffering from irritation, chronic or irreversible tissue damage or narcosis to a degree sufficient to increase the likelihood of accidental injury, impair self-rescue or materially reduce worker efficiency.

Stretcher—See litter; the terms are used interchangeably.

Stopper Knot—A knot, usually a figure 8 or overhand, tied in the end of a rope to prevent equipment from sliding off the end of the rope or to warn a person on rappel that the end of the rope has been reached.

Support (Cold) Zone—The area outside of the Exclusion (Hot) Zone in which PPE or other safety equipment is not required.

—T—

Tachypnea—Excessively rapid respiration.

Tag Line—A line tied to entrants or rescuers in a space and dragged behind them to aid in locating them in the event of an emergency. It is also a line attached to the stretcher to prevent it from spinning on a vertical lift, to pull it away from obstacles or on a high line to move the stretcher from one side to the other.

Tandem Prusik Belay—Two three-wrap Prusik hitches on the belay line and connected to the belay anchor through a load-releasing hitch or device. The Prusik loops are slightly different lengths, so that the Prusik hitches are positioned about three inches apart.

Technical Use—Formerly Light Use, an equipment designation in *NFPA Standard 1983 Life Safety Rope and Equipment for Emergency Services.* Technical Use has the middle performance requirements of the three designations: General, Technical and Escape.

Technician Level—The highest of the three levels of operational capability. Technician level personnel and organizations are trained to perform hazard identification, use equipment, perform advanced technical rescue techniques, supervise other responders and manage technical search and rescue incidents.

TLV®—Threshold Limit Value is a registered trademark of the American Conference of Governmental Industrial Hygienists (ACGIH) and refers to the airborne concentration of substances to which it is believed that nearly all workers can be repeatedly exposed eight hours a day over a working lifetime without adverse affect.

Toxic—Harmful, deadly or poisonous when the level of contaminants exceeds the OSHA Permissible Exposure Limits (PELs), ACGIH Threshold Limit Values (TLV®s) or NIOSH Recommended Exposure Levels (REL).

Tripod

Tri-axial Loading—When the force on a carabiner is exerted in three directions rather than along the main axis. This shifts the load away from the spine of a D-shaped carabiner with a reduction in its functional strength.

Tripod—A portable anchor device with three legs used as either a high point anchor for access or retrieval or as a high directional.

—U—

UEL/UFL—Upper Explosive Limit/Upper Flammable Limit is the richest mixture of fuel and oxygen that will permit ignition.

Unified Command—In ICS, unified command is a unified team effort that allows all agencies with responsibility for the incident, either geographical or functional, to manage an incident by establishing a common set of incident objectives and strategies. This is accomplished without losing or abdicating agency authority, responsibility or accountability.

—V—

Vapor Density—The weight of a given volume of gas or vapor compared to an equal volume of air at the same temperature and pressure. Air has a vapor density of 1. Gases with a vapor density of less than 1 are lighter than air and will rise. Gases with a vapor density of greater than 1 are heavier than air and will sink.

Vector—A force that has a magnitude and a direction.

Vector—An organism that carries pathogens, such as a mosquito or a rat.

Ventilation Control—The control of atmospheric hazards by introducing clean, breathable air into to the confined space.

Vertical—An evacuation that is essentially free hanging.

—W—

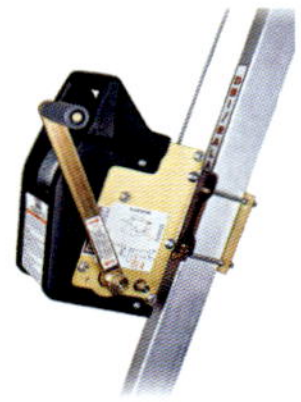

Winch

Wristlets

Winch—A device for lifting or lowering a load with an integral cable or line and with gearing to provide a mechanical advantage when lifting or lowering.

Working Load—The maximum load approved for a rope by the manufacturer. The working load is a percentage of the minimum breaking strength of the rope and varies depending on the type of construction.

Wristlets—Straps designed to be placed around the wrists or ankles of a person to allow lifting, lowering or dragging.

—Z—

Z-Rig—Another term for a 3 to 1 mechanical advantage system from the Z shape the rope makes when the system is rigged.

INDEX

NOTES